AETHER T

AN ALTERNATIVE PHYSIC
AETHER

by
Reginald F Norgan

June 2010

GRETA PUBLICATIONS

ii

Published in Great Britain in 2010 by Greta Publications.

First edition

ISBN: 978-0-9566023-0-5

Printed and bound in Great Britain by **think***ink*

To

Ron and Auria Barker
and Cheryl

Without whose encouragement
this book would not have been written.

iv

Contents

Introduction 1

Measuring the Speed of Light 22

The Nature of Time 44

The Michelson-Morley Experiment 55

Special Relativity 70

The Electric Field 82

The Aether Theory of Velocity Effects 99

The Twins Paradox 109

Magnetism 122

Electric Wave Transmission 136

Gravity 144

Space Expansion 166

Appendices 180

.

(**Newton, 1687**)   Absolute Space in its own nature, without relation to anything external always remains similar and immovable.

(**Faraday, 1830**)   I cannot conceive curved lines of force without the conditions of a physical existance in that intermediate space.

(**Maxwell, 1876**)   In speaking of the energy of the field, however, I wish to be understood literally. All energy is the same as mechanical energy, whether it exists in the form of motion or in that of elasticity, or in any other form. The energy in electromagnetic phenomena is mechanical energy.

(**Lorenz, 1906**)   I cannot but regard the ether, which can be the seat of an electromagnetic field with its energy and its vibrations, as endowed with a certain degree of substantiality, however different it may be from ordinary matter.

(**Einstein, 1928**)   According to the general theory of relativity space without ether is unthinkable, for in such space there not only would be no propagation of light but also no possibility of existance for standards of space and time.

Chapter 1

Introduction

During a short hiatus in my career I looked for some fresh interest with which to fill my newly available free time. Having always had some interest in physics I decided to investigate Einstein's Special Theory of Relativity. At that time I knew practically nothing of the theory or of its predictions. Einstein himself seemed to me like some god of physics, head and shoulders above all others, and I doubted whether I possessed the intelligence to understand any theory which he had created. To my great surprise I found his theory relatively easy to understand and furthermore found the predictions of the theory - length contraction, time dilation and mass increase - easy to accept. Nevertheles, I did find a serious problem with Relativity. I found myself totally unable to accept the strange Universe that Einstein's theory had created.

Special Relativity (SR)
In his theory Einstein postulates that the laws of physics are identical in all inertial reference frames (IRFs) and that no IRF is any more special than any other. Crudely speak-

ing we may take an observer on Earth to exist in an IRF. Similarly an observer inhabiting the fictional planet Zeno orbiting a distant star and travelling through Space at a constant relative velocity of a thousand kilometres per second with respect to our Sun, would also exist within an IRF. Einstein's postulates state that both of these observers (as with all IRF observers) find the laws of physics to be identical, despite the large relative velocity between the two observers.

For example, the universal constant the speed of light, must be measured to be of equal magnitude by all observers. So, if the planet Zeno emitted a beam of light which crossed Space to shine upon Earth the Zeno based observer would measure the velocity of the departing beam to be 299,792km/s while the Earth based observer measuring the same beam as it arrived on Earth would also determine its velocity to be 299,792km/s. According to Einstein both these measurements, by the Zeno and the Earth based observer, remain identical whatever the relative velocity of the two planets. This statement remains true even if the relative velocity were as high as 299,000km/s.

A simple analogy might be that of a boy standing in the back of a truck moving along a road at 49km/hr who then throws a ball forwards at a velocity of 50km/hr. The ball is then caught by another boy standing in the road ahead. The stationary boy finds the velocity of the ball with respect to himself to be 99km/hr, which is what we would expect, being the addition of the throwing velocity and the truck velocity. But on the basis advocated by Einstein the stationary boy would find the velocity of the ball to be 50km/hr rather than 99km/hr. Hence Einstein's postulates violate the law of addition and subtraction when applied to the velocities of light and matter.

One might try to explain this paradox by surmising that Einstein really stated that the *measurement* of the speed of light is constant but the true velocity is rather different. This would imply that our units of distance and time are unwittingly modified in some manner so as to create the difference between the measured and the true figure. But if this had been Einstein's true meaning then he would actually have been advocating an Aether based theory. No! Einstein firmly states that the speed of light *is* identical in all IRFs rather than merely appears to be identical.

I have to say that I totally failed to create a picture in my own mind of a Universe in which velocities do not obey the laws of addition and subtraction. Yes, I am rather old-fashioned in that respect. I firmly believe that if you can't form a picture of the explanation then that explanation is probably wrong.

But the situation is even more bizzare. Einstein states that an observer's measure of a matter body describes the only 'reality' of the body that we may obtain; despite the fact that all observers, moving at entirely different velocities relative to the body, make different measurements of that body. Consequently each matter body possesses an infinity of different 'realities', and no one of them is of any more correct than any other. Einstein, therefore, totally dispenses with a physical reality independent of observation. From this it must be deduced that material bodies cannot exist unless they are being observed, for otherwise they could have no specific size or mass. The logical outcome of Einstein's theory is therefore an empty Universe.

As a requirement of Einstein's SR postulates, it follows that Space cannot be a physical substance relative to which the propagation velocity of light is determined, otherwise the laws of physics would relate to that one Space substance rather than to Einstein's multiple IRFs. And yet, up to the publishing of Einstein's theory in 1905, the great majority of physicists thoroughly accepted that Space was indeed a physical substance, called the luminiferous Aether or Ether. Newton, Maxwell and Lorenz are just three of the eminent phycisists who accepted the existance of the Aether substance.

So why do I instinctively believe that Space must be an Aether substance and consequently find it impossible to accept Special Relativity. As in all things it is a matter of judgement. One gathers the facts appropriate to both sides of the argument and applies them to the weighing machine of the mind. There are many different reasons to support the Aether argument but I believe the three main ones to be:-

- *Celestial bodies such as stars are separated from each other by varying distances.*
  Something must determine this variable degree of separation, which is exact for a particular instant in time. In simple terms the celestial bodies must be positioned within a physical Space matrix in a similar manner to the positioning of currants in a fruit-cake.

- *The velocity of light is 299,792,458 m/s.*
  This very precise propagation velocity of light through Space must be determined by the physical properties of the substance through which it moves. This concept is analogous to the determination of acoustic velocity through a material substance as a function of the stiffness and the density of that substance.

- *Matter must possess specific dimensions and mass independent of observation.*

Why SR is Widely Accepted

On the other hand, why is it that modern physicists ignore the above points, along with their natural intuitions, and choose not to believe in the Aether.

One answer is because they believe the Special Theory of Relativity to be correct - and that theory denies the existance of the Aether. In turn, they believe in the Special Theory because its predictions have been found to agree with observation. Thousands of experiments and observations designed to test various aspects of the theory have been undertaken since 1905 and all (bar one, the Twins Paradox) have been accepted as proving the theory correct. Experimental evidence appears to be entirely (almost) on the side of the Special Theory.

A further answer might be the fact that Einstein has evolved into the God of Physics. Einstein is known by every household in the land. He is the epitome of the scientist. His picture with his unruly hair is known by everyone. Indeed his face is more instantly recognised than that of Charles Darwin, the only other great scientist of popular renown. The physics community has chosen to go along with the deification of Einstein because of the benefits that it brings. Anything which elevates the profile of science in the public mind eventually filters through to the politicians. And the politicians hold the public purse strings which fund scientific projects and the salaries of the professional academics. Do not get me wrong. I am not criticising the scientific community for using Einstein as a tool to prize open the public coffers. We all live in a pragmatic world in which we have to fight for our share. But the cost in the deification of Einstein is that his theories are, to an extent, placed beyond criticism.

A further mechanism in operation which biases judgement on the side of Special and General Relativity is that no-one likes to admit that they are wrong. Young Phds who have only recently mastered the intricacies of Special and General Relativity are naturally very proud of their achievements. They have gained a position of status in the world which few others have attained. Naturally they do not take kindly to anyone who tells them that, unfortunately, they have it all wrong. Professors who have accepted Einstein's theories for all their academic lives are even less likely to seriously consider any criticism of their beliefs.

Perhaps another reason why physicists believe in Special Relativity is that they are reluctant to accept the Aether when it has never been directly detected by humans employing even our most complex and accurate instruments. They hold to a philosophy which states that if something is impossible to detect then it cannot exist - and on this basis the Aether cannot exist. It is an easy philosophy to intuitively accept but it does assume the total power of matter (of which humans and their instruments are constructed) to detect the more fundamental components of which matter is itself constructed.

Objections to the Aether Hypothesis
The opponents of the Aether hypothesis put forward two major objections. They reasonably argue that if the Earth is moving through the Aether - which as the Earth is orbiting the Sun at 30km/s must be the case - then the measured speed of light should vary by at least that amount. The famous Michelson-Morley (M-M) experiment was designed precisely to measure this difference. But as Michelson and Morley measured no difference whatever in the velocity of light in any direction through the heavens the opponents of the Aether hypothesis immediately take this result as proof

that the Aether did not exist.

But measuring the one way speed of light is not possible as this book explains. The necessary two way measure averages a fast transit in one direction with a slow transit in the opposite direction, such that the measured average velocity is not greatly different whatever the Aether velocity of the observer. For the Earth's orbital velocity the difference is just one part in 100 million. Nevertheless, the Michelson and Morley apparatus would have detected this minute effect had it existed. They did not detect it because their apparatus was length contracted to a degree which exactly canceled out the expected difference. For Aether theory states that all matter suffers length contraction in the direction of its Aether motion as a function of Aether velocity.

Opponents of the Aether hypothesis suggest that the fact that the length contraction of the M-M apparatus is predicted to *exactly* cancel the expected velocity measurement difference requires too high a degree of probability to be believable or acceptable. But in this they misunderstand the nature of matter. This book explains that matter contracts because the electric fields of the nucleii of atoms (which determine the orbits of the electrons and hence the dimensions of the atom) are subject to exactly the same two way propagation velocity effects as the light beam being measured. Thus it cannot be otherwise that the out and return time for a light beam to traverse a matter determined distance (eg. an arm of the M-M apparatus) must be a constant, whatever its Aether velocity.

In fact it is completely immaterial which type of experiment we employ in an attempt to measure or discover our Aether velocity. It is the case that all of our measuring instruments of whatever type are distorted, both in length and in time, by their own Aether velocity. That distortion in our instrumentation inhibits us from ever knowing the degree of that distortion. Hence the Aether hypothesis leads directly to the conclusion that the Aether cannot be detected by any *local* experiment whatsoever. Put in another way, the laws of physics appear invariant in all inertial reference frames.

But that is not to say that the Aether completely denies its own existance. Indeed, the Cosmic Background Microwave Radiation, sourced by bodies many billions of years ago which effectively marked the position of the Aether at that time, shows via the Doppler effect that the Earth is moving at a velocity of approximately 360km/s relative to those sources, and hence relative to the Aether.

The second major objection to the Aether hypothesis is that if Space is a substance then how does the Sun, for example, move through the Aether substance at a velocity of 360km/s without friction or turbulence.

It must be the case that one substance cannot freely move through another substance without loss of velocity. Either one or the other must be insubstantial. As the original objectors intuitively believed that matter was substantial then it followed that Space could not be substantial and so could not be an Aether. On the other hand, if Space is taken to be a substance then mass must of necessity be insubstantial. Consequently believers in the Aether are forced to closely examine matter and mass in order to understand their fundamental structure. This requirement is crucial to a successful Aether theory.

The strongest clue to the real nature of mass lies in the fact that mass and radiant energy are interchangeable through the famous equation, $E = mc^2$. If radiant energy consists of electric waves then it follows that mass, too, is likely to be constructed of electric waves. The difference between mass and energy most likely lies in the different geometries of their electric waves - a straight line movement for energy and some form of rotatory movement for matter. On this hypothesis mass is no more substantial than a ripple in the Aether, albeit that ripple may be of a very complex form. Hence every massive object, however large, is merely an infinite complex of electric ripples superimposed upon the Aether.

At this stage of our knowledge it is not possible to fully describe the construction of fundamental mass particles (FMPs). Nevertheless, the simple hypothesis that FMPs are constructed of electric waves leads to various important and far-reaching conclusions.

Relative and Absolute Velocity

Since 1905 many scientists have been unhappy with the loss of the Aether to science. They believed that the Aether hypothesis leads to more acceptable explanations of many different and varied physical phenomena than are currently provided by non-Aether physics. In an attempt to restore the Aether to physics several physicists have tried to create a theory of relative velocity effects based upon postulates in keeping with the Aether hypothesis. Herbert Ives, Geoffrey Builder and S. J. Prokhovnik come to mind among those who made the attempt. It has to be said that none of these gentlemen were successful in their efforts. After all it could never be an easy task to generate a new theory which makes identical predictions to those of the original theory but is based upon completely opposing postulates.

It must be the case that the Aether hypothesis is intuitively expected to lead to theory of velocity effects which is a function of *absolute* velocity, where that function may or may not be similar to that derived by Special Relativity. In fact, the absolute velocity functions of the Aether

Theory of Velocity Effects (ATVE) are identical to the SR relative velocity functions. But it is the case that within local experiments we do not observe any phenomenon which is a function of absolute Aether velocity. Without exception, the observed effects of countless experiments are found, instead, to be functions of *relative* velocity - that is the velocity between the IRF of the observed body and the IRF of the observer. Thus Aether theory, at this stage at least, does not predict what we observe.

As we can never know the absolute velocity of any body it would appear that a theory of absolute velocity effects is of no value whatever. But strangely, the conversion from the Aether theory of absolute velocity effects to a theory of relative velocity effects is an extremely simple and straightforward one. All that is required is the application of the standard Lorenz Velocity Transform equation. By far the most difficult part of the ATVE lies in the generation of a theory of absolute velocity effects in the first place.

The predictions of the new theory - The Aether Theory of Velocity Effects in its relative velocity form - being identical to those of Special Relativity naturally agree with all experiments and observations on relative velocity effects made to date. Having said that, the two theories are far from identical in their application. For Special Relativity has the crippling disadvantage that its postulates require the theory to be applied solely to observations between inertial reference frames and hence between inertial bodies. In practice material bodies are never totally inertial - Earth based laboratories in particular are not inertial due to the rotation and the solar orbit of the Earth. Consequently Special Relativity cannot be employed in any real observation or experiment whatever.

On the other-hand the ATVE does not require either the

observed body or the observer to be inertial. The ATVE therefore has no difficulty in predicting the observations of the travelling twin of the stationary twin in the famous Twins Paradox type of experiment. In this particular circumstance the ATVE predicts time *contraction* - a prediction of which SR is totally incapable. The Twins Paradox experiment is a most significant experiment in highlighting the fundamental differences between Special Relativity and the ATVE such that this book devotes an entire chapter to it.

One More Change

It is the case that much of physics must differ considerably when based upon the acceptance of Space as a physical substance relative to the Einsteinian understanding that Space is some complex combination of Space and Time represented only by an infinity of overlapping IRFs. However, Aether physics requires one further fundamental change to the accepted understanding of physics and its laws. *Aether physics denies that magnetism is a fundamental force.* Without that crucial change the Aether Theory of Velocity Effects cannot be constructed. It might well be said that Modern physics and Aether physics diverge from this difference in the understanding of magnetism and take two entirely different paths.

If it is assumed from the beginning that magnetism is a fundamental force then from that point on subsequent physics must always accommodate itself, however awkwardly, to that initial concept. If on the other hand magnetism is not considered a fundamental force then future physics never finds the concept to be necessary, as this book demonstrates.

The force exerted by one magnet upon another does not immediately appear at first sight to be a derived manifestation of any other known force. Consequently magnetism was originally considered to be a distinct and separate force. James Clerk Maxwell later decided that the magnetic force and the electric force were actually manifestations of each other when observed from a relatively moving IRF at various velocities. According to Maxwell a force was either magnetic, electric or a combination of both forces depending upon the situation of the observer. Thus, in Modern Physics the magnetic effect is currently considered to be of equal significance to the electric effect in the theory of electromagnetism.

Aether physics takes a radically different view. Firstly it considers the electric potential to be physically represented by the Aether as a scalar potential at each and every point in the Universe. The Aether substance may be considered to be formed of identical Aether units, termed Aethons, analagous to the atoms of a material substance, where each Aethon supports just the electric and the gravitational scalar potentials. Aether theory suggests that the electric potential is represented by Aether pressure and the gravitational potential by Aether density.

Each Aethon effectively acts as a source of potential whereby its potential difference to its immediate neighbours is passed on to them via intimate contact at the speed of light. By this mechanism the influence of each Aethon is found to diminish with inverse distance. But simultaneously each Aethon is also a recipient of the potentials passed on by *its* neighbours. This is the Aether Transmission Mechanism.

The absence of magnetism as a fundamental physical phenomenon naturally has a significant effect upon physics. For example, electromagnetic waves are now simply electric waves.

The famous Maxwell equations, which describe the interaction between the electric and the magnetic force, no longer have any meaning.

As a further effect the equation which describes the electric potential field surrounding a moving charge differs to that employed in electromagnetism as the new electric field must now describe the total effects previously ascribed to the electric and the magnetic forces. In fact the new electric field is more simply derived than before as the concept of charge volume and the Lienard-Weichardt potentials are totally discarded by Aether theory. Instead the electric field is simply derived from the consideration that a charge is a source of greatly elevated electric potential occupying an effective point relative to the dimensions of the field where the surrounding potential field is generated by the Aether Transmission Mechanism.

The electric field is transmitted outwards away from the charge through and relative to the Aether but as the charge moves through the Aether it also moves through its own field. Thus the electric field surrounding a moving charge is asymmetric about the charge.

It is at this point that Aether Physics needs to employ the Two-Way Maxim. The Two-Way maxim is originally derived from the conclusion that it is impossible to measure the one-way velocity of light - but the maxim is of a far more general nature. The maxim states that meaningful communication between bodies must be two-way. For example, if charge A affects and moves charge B then charge B will reflect its movement back to charge A via its own electric field - similarly via the gravitational fields of two massive bodies. Communication may be by any means and at any mix of velocities.

The electric potential field surrounding a charge is therefore effectively 'observed' via its action upon a second charge, which in turn communicates back to the original charge at the speed of light via its own field. The 'observed' electric potential field is now symmetric about the charge but, as expected, the Aether equation describing the field is rather different to the Maxwell equation.

Not only does magnetism not exist in Aether theory but the creation and the form of the electric field is also very different.

The Lorenz Transforms

The dimensions of the observed electric potential field sourced from a charge moving through the Aether may be compared to the dimensions of an electric field sourced from a stationary charge, where the observer is himself stationary in the Aether. As the field is 'observed' it is effectively measured by an out and return communication at the speed of light to any chosen point of the field.

The dimension of the observed field in the direction of movement of the charge is found to be contracted relative to the two transverse dimensions as a function of Aether velocity. The two transverse dimensions are also found to be contracted relative to the stationary field, and to an equal degree.

But what of the Time dimension.

This book devote a whole chapter to the understanding of the nature of Time. But time cannot be understood without a prior understanding of the nature of mass, for the variable nature of matter velocity otherwise seems to preclude any conclusion. However, if matter is just a complex of electric waves as previously hypothesized then, viewed fundamentally, all electric waves move at the propagation velocity of the Aether. Thus, time is perfectly analagous to distance moved.

So Time is not a separate dimension and neither is it

universally constant. Time exists, but embedded in the Aether within the local propagation velocity, which varies from point to poin in Space. For an observer occupying an inertial reference frame time is two-way IRF distance moved at the speed of light. As distance varies between IRFs then so our concept of time varies.

These relationships between distance and time with respect to different IRFs aregiven by the famous Lorenz Transforms. But they are immediately only relevant to the electric potential field from which they are calculated and then only between the Aether and an IRF. However, by the assistance of the eminent physicist, J. S. Bell, it is shown that the length contraction and time dilation effects of the electric field lead directly, via the construction of the atom, to an identical effect upon matter in general. Consequently the Lorenz Transforms are equally applicable to bulk matter as to the electric field.

The further Lorenz Transform equations relating to mass, velocity and the inter-relationship of mass and energy are simply derived from those of distance and time. Thus the various dimensions of matter are a function of absolute Aether velocity. But unlike Special Relativity these dimensions are independent of observation. Therefore material bodies in Aether Theory possess their own dimensions. They are real.

Thus we arrive at the Aether Theory of Absolute Velocity Effects. This theory can be readily converted to the Aether Theory of Relative Velocity Effects as previously mentioned.

Magnetism

Magnetic effects are generally associated with current carrying wires.

Although not at all obvious the magnetic effect is actually an electric effect. It is the net effect of two opposite polarity electric fields situated at the same point, where one field is slightly different from the other. The difference in the two fields is caused by a small difference in the Aether velocities of the two charges as described by the new electric field equation. Therefore the net difference is a function of their relative velocity.

This situation is most commonly experienced in the atom where the negative fields of the electron cancel the postive field of the nucleous. In bulk matter the electrons, as they orbit the nucleii of the constituent atoms, move equally in all directions and so the average net effect is zero. However, in a metal some electrons are free to move through the material such that a net directional current flows when a potential difference is applied. The relative velocity of the current electrons leads to a net electric effect, but additionally it requires that an 'observer' charge must also be moving relative to the wire before the effect is experienced.

The observed magnetic forces and the phenomenon of induction are all explained by this process.

Action at a Distance
In the absence of the Aether to act as a carrier of potential across Space, Modern Physics finds great difficulty in explaining how a body at one point in Space affects a second body some distance away. These effects over distance are called forces.

Modern Physics states that forces are mediated across Space by force carrier particles called bosons, where each type of force employs a different boson.

For example, the virtual photon is proposed as the carrier for the electromagnetic force. But just consider all that the carrier photon needs to do. If two separated charges have the same polarity the virtual photons causes a mutual repulsion - if of opposite polarity they cause an attraction. Alternatively, if the two bodies are magnetic poles (accepted by Modern Physics) the carrier photon causes a repulsion when the two poles are of like kind and an attraction if they are different. But the carrier photon does not cause acceleration between a charge and a magnetic pole despite the fact that it causes accelerations between charges and between poles.

As a photon only carries two pieces of information, direction and energy, it does not have the capacity to carry out all of these requirements. Firstly a charge would need to make the decision whether the creation of a virtual photon is required and if so at what energy and in which direction should it be emitted. The reception of that virtual photon by a second and distant charge must somehow convert that reception into an acceleration of a specific magnitude in either a direction towards the source of the photon or in the opposite direction. How can this all be done? One can continue indefinitely with these awkward questions but it is obvious that the whole concept of force carrier bosuns is totally flawed.

The proposed gravity force bosun, the graviton, has never been discovered and furthermore does not stand up to scientific logic.

Comparatively speaking the Aether explanation of Action at a Distance is simplicity itself. As previously stated the Aether supports the electric and the gravitational potential at all points in Space. Points of elevated potential such as charges and massive bodies create surrounding fields of diminishing potential via the Aether Transmission Mechanism. This mechanism ensures that the individual fields of every source, situated anywhere in the Universe, superposition upon each other to create an ambient field. Each fundamental matter particle then responds to its local ambient field by accelerating at a rate in proportion to the potential gradient there and in the direction of that gradient. Oppositely charged particles accelerate in opposite directions in the same electric gradient.

The complexity of the Aether solution to Action at a Distance lies mostly within the hypothesis of the construction of a Fundamental Matter Particle which somehow must explain the internal geometry causing the particle to move through the Aether at a particular fraction of the speed of light - and then how that geometry is modified by the super-position of local potential gradients upon the internal potentials of the particle.

Gravity and Cosmic Expansion

Sir Isaac Newton's theory of gravity explained that mass generates a gravitational field where the strength of the field is proportional to the amount of source mass. The gravitational field causes mass situated within that field to accelerate towards the source of the field at a rate which diminishes with the inverse square of distance from the source. Thus the field, in theory, extends to infinity.

Einstein improved upon Newton's theory with his General Theory of Relativity in which he demonstrated that light is also affected by the gravitational field, in that the speed of light diminishes the nearer the source mass is approached. The gradient of the speed of light so produced causes light travelling across the gradient to bend towards the region of slower speed of light. But in order to explain these effects Einstein decided to create a new form of Space called Space-Time. Space-Time is a complex mathematical construction employing an imaginary time dimension which is impossible to envisage - and furthermore is scientifically unsound.

But neither Newton or Einstein explained to the smallest degree how a source mass causes a gravitational field. Aether theory takes a step in that direction.

Aether theory states that the gravitational potential is simply the local speed of light, and that the speed of light is caused to slow close to mass. The Aether Transmission Mechanism creates a field of speed of light difference (relative to an arbitrary distant point) which diminishes with inverse distance from the mass source. The gradient of this field causes light rays to bend when crossing the gradient. This bending is the essence of gravity. The action of the speed of light gradient upon the internal propagating electric potentials of a fundamental matter particle is not so obvious but is expected to modify the internal geometry which determines the Aether velocity of the particle.

Thus gravity is simply an effect of refraction in Space.

It is found that stationary matter accelerates at one half the rate of light at the same speed of light gradient. This is theoretically justified to an extent by the hypothesis of the structure of Fundamental Mass Particles. The factor of one half diminishes to zero as the velocity of the mass particle approaches luminal velocity in the direction of the accelerating gradient.

The slowing of light close to mass is explained by the revolutionary postulate that mass generates new Aether. The new Aether then diffuses outwards through Space to create a field of excess Aether density diminishing with inverse distance. This increase in Aether density - with no increase in Aether pressure - causes the propagation velocity to diminish in a manner analogous to the determination of the propagation velocity within bulk matter. This variation in Aether density in the region of mass creates a non-Euclidean Aether.

The Aether theory of gravity is intrinsically modified from the Newtonian form as the outward diffusion of new Aether is affected by the general expansion of Space observed in the phenomenon of Cosmic Expansion, with the result that the excess Aether density falls off more rapidly with distance than the inverse square law. Thus an additional degree of acceleration is generated. Nevertheless the effect is very small and therefore only noticeable at very low levels of acceleration. Measurements of the deceleration of the two Pioneer spacecraft, which are travelling out of the solar system, exhibit this effect. Non-Aether physics has so far failed to provide an alternative explanation.

It is also possible that this same mechanism might explain the higher than expected gravitational acceleration observed in the outer regions of galaxies. This is currently believed to be caused by the presence of Dark Matter. But to suggest, as Modern Physics does, that there exists an undetected dark material of ten times the abundance of visible matter in order to explain a gravitational anomaly of only $1.2 \times 10^{-10} m/s^2$ seems to be overkill to an excessive degree.

If Space is an Aether substance then, as Space is observed to be expanding, the expansion can only result from the creation of new Aether substance. The Aether Theory of Gravity postulates that mass creates new Aether, but it should be remembered that mass is not a substance in itself but merely a volume of Aether in which intense electric potentials move in some complex geometry. Thus it is the Aether internal to the mass particle which generates the new Aether. Aether theory considers that the new Aether necessary for the observed rate of cosmic expansion is generated by Aether of very low matter density - in other words the Aether of outer Space. New Aether generated by mass, albeit at a very much higher rate, is expected to contribute very little to cosmic expansion due to the low average density of matter in the Universe.

Thus gravity and cosmic expansion are seen to be related. Gravity is a result of the gradient of a specific localised flow of new Aether while the more uniform creation of new Aether by empty Space creates cosmic expansion.

Wave Transmission through the Aether

The Huygens-Fresnell explanation of the reflection, refraction and diffraction of electric waves is based upon a Space transmission mechanism which is practically identical to the Aether Transmission Mechanism. Thus we have the bizzare situation where one part of modern physics is effectively based upon Aether physics while other parts, Special and General Relativity, are based upon a denial of the Aether.

Aether physics only needs to supply one absent explanation - why we experience only forward wave transmission.

The Provenance of the Theories

All of the theories described in this book, where they are not ascribed to other authors in the text, are entirely the work and invention of the author. It may be that others have duplicated these theories either in part or wholly, either prior to my personal invention or indeed after. However, where that may be the case I have had no knowledge of the work of others which might otherwise have been constructive in the generation of my own theories. It is of course the case that I have seen many attempts by others to create theories to explain some of the phenomena which I myself attempt to explain. But in all of these cases there have been fatal flaws which made their theories untenable.

Chapter 2

Measuring the Speed of Light

It is possibly true to say that Michelson and Morley's attempt to measure the speed of light marked the beginning of the age of Relativity. Although Relativity did not officially begin until the publishing of Albert Einsteins famous paper, the Theory of Special Relativity, in 1905.

The exact determination of the value of the universal constant, the speed of light, is extremely important to science; but an understanding of the systemic problems which directly arise from the attempt to measure it also tells us much about the nature of Space, Time and Matter.

It is obvious to all that light is generated by a source of some kind or other, eg. the Sun, a light bulb, etc, which then illuminates objects at some distance from the source. For example, the Sun illuminates all of the planets of the solar system, including the Earth, over distances of billions of kilometres of Space. It is logical to assume that light travels the intervening distance between source and illuminated object at a certain velocity, just as sound travels from a distant source to our ears at the propagation velocity of 330m/s - a figure known for many centuries. But prior to the 17th century the velocity of light was entirely unknown. In truth, prior to this time the velocity of light was believed to be infinite, as objects appeared to be illuminated as soon as the light source was uncovered, irrespective of how little or how great the separation distance between object and source.

The speed of sound can be readily judged by timing the delay between the flash of a distant cannon and the receipt of the sound of the explosion. If the cannon were situated one mile away the delay would be roughly 5 secs. But the difficulty in visually assessing the speed of light may be realised by reflecting a light beam from a mirror placed 1 km from the source. The velocity of light is today known to be 298,792 kms per second so the round trip time from light source to mirror and back again is a mere $7\mu s$. The human brain is quite incapable of detecting such a short time period, and no simple timer or clock can measure such a short interval. Therefore, quite understandably, it was assumed that the time delay was zero and hence the speed of light was infinite.

Early Measurements

Galileo was the first person to propose a method of measuring the speed of light. He suggested that a lantern be uncovered which was visible to a second person placed at the maximum distance at which the lantern could be observed. On seeing the light from the distant lantern the second man uncovered his own lantern as rapidly as possible. The first lantern man then assessed the elapsed time between the uncovering of his lantern and his reception of the returning light from the distant lantern. Unfortunately, at that time Galileo was blind and under house arrest for

publicly supporting the Copernican idea that the Earth revolved around the Sun, rather than the popular belief that
it was the other way round. Consequently Galileo could
not carry out his proposed experiment. The experiment was
eventually undertaken 25 years later after his death.

Galileo's experiment was repeated many times over with
the separation distance between the two lanterns continuously increased. The round trip delay time did not seem to
be affected by the changing separation distance as the human reaction times involved - being indepedent of the separation distance - were the main cause of the delay time. The
experimenters eventually came to the rather weak conclusion that the speed of light was somewhere between 3 km/s
and infinity.

Although the experimenters were a very long way from
the true figure the actual principle of Gallileo's experiment
was not at fault. Their major difficulty was a technical one,
in that they employed a relatively slow human-being to both
switch and detect the beam and to ascertain the time delay.
The speed of light was far too fast and the course length
too short for the measurement capability of their human
'apparatus'.

The first reasonably accurate measurement of the speed
of light was made in 1676 by the Danish astronomer, Ole
Roemer, who worked at the Paris Observatory and at Copen-
hagen. Roemer employed Cassini's measurements which had
been made in a careful study of Io - one of the moons
of Jupiter. Io orbits Jupiter with a precise period and is
eclipsed by Jupiter at regular intervals. The time between
eclipses should therefore also have occurred at regular intervals, but instead Cassini found that, over a period of several
months, the eclipse time came later and later until eventually they began to sequentially speed up again over a further

period of several months. Roemer eventually realized that
the periods of slowing and speeding up of Io's orbits occurred
over a period of exactly twelve months. The effect therefore
seemed to be something to do with the Earth's orbit around
the Sun.

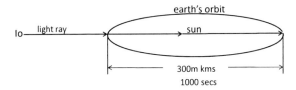

Figure 2.1: Roemer's Measurement

Roemer realized that if light traveled at a finite rather
than an infinite velocity through Space it would take a longer
time for the light from Io to reach Earth when Earth was at
the point in its orbit the furthest away from Io than when
Earth was at its nearest point six months later. The inte-
grated diminution in the measured eclipse times of Io over a
period of six months is therefore the time taken for light to
cross the diameter of the Earth's solar orbit. Knowing the
orbital traverse time from Cassini's observations combined
with a measure of the distance from the Earth to the Sun
enables the speed of light to be simply calculated. In fact
Roemer never made the calculation himself - it was later
made by Christiaan Huygens using Roemer's and Cassini's
data. However Huygens measure of the speed of light was
rather inaccurate as the diameter of the Sun's orbit was not

well known at the time. A more accurate figure for the
distance from the Earth to the Sun became available later
which gave a measure of the speed of light of 214,000 km/s.
This figure was considerably greater than previous estimates
and much nearer the true value of 299,792 km/s. The speed
of light was indeed very, very fast - but it was not infinite
as previously thought.

The next improvement in accuracy came about in 1727
when an astronomer named James Bradley managed to measure the speed of light - although once again in a way purely
incidental to his intentions. Bradley was attempting to map
the exact position of the stars in the heavens using the lat-
est advances in telescopes but he found that the position of
the stars did not remain constant over the period of a year.
In fact their changing position in the Heavens described an
ellipse. He finally realized that this was due to the velocity
of the Earth in its orbit round the Sun combining at near
right angles with the light from the star.

An analogy might be a man walking around in a circle
at say, 3 km/hr, in the rain which is falling down vertically
at 30 km/hr. Due to the motion of the walking man the
rain always appears to be striking him from the front at an
angle of approximately 6 deg., no matter in which direction
he walks. When he walks north the rain appears to be at
an angle of 6 deg. from the north and when he walks south
the rain then appears to be coming from an equal southerly
angle. Due to his walking speed the man observes a total
aberration of 12 deg. in the direction of the rain.

Similarly when a star is in a position in the night sky
at 90 deg. to the orbit of the Earth, the velocity of the
Earth and the velocity of light make a right angled triangle
with one side about 10,000 times longer than the other side.
The small angle of this triangle - about 21 seconds of arc

- created a degree of apparent deflection of the position of the star. The deflection angle can be obtained by measuring the angular position of the star at six-monthly intervals on opposite points of the Earth's orbit. Knowing the Earth's orbital velocity and the angle of deflection gave Bradley a measure of the speed of light of 297,000 km/s. Bradley's measure was accurate to within one percent of the modern figure.

The first person to measure the speed of light entirely on Earth, rather than through astronomical observation, was a Frenchman named H. L. Fizeau. Fizeau used the same principle as Galileo, but instead of using a second lantern man he employed a mirror placed some 15 kms from the light source which reflected the light back to the source. Fizeau also used a more powerful light source than the Galileo experimenters, concentrated by a system of lenses to pass between the teeth of a revolving wheel on the way to the distant mirror. The cogs of the rotating wheel served to chop the light source into a series of pulses - rather like covering and uncovering a lantern at an extremely rapid rate. The returning reflected beam was once again made to pass through the teeth of the revolving wheel to be observed on the other side.

Initially the wheel was revolved slowly such that the returning beam passed through the same slot from which it had emerged. Then the rotation rate was increased until eventually the next cog on the wheel moved into the line of the returning beam and prevented it from passing through. At this point Fizeau had all the information available to calculate the speed of light. He knew the time for one tooth to pass the beam from the rotation rate of the wheel and the number of teeth around it. He also knew the distance to the mirror and back. Dividing the distance by the time gave a figure for the speed of light. The accuracy of Fizeau's experiment was such that it gave a measurement within 1500 km/s of the true value.

It began to seem that measuring the speed of light was no longer a difficult problem. But this was far from the case.

The Concept of the Aether

The attempt to measure the speed of light gives rise to a most crucial philosophical question - exactly what is it that determines the precise value of this Universal parameter.

At the time of the early measurements there were two different theories as to the nature of light. One theory was that light was made of tiny particles. In this case the speed of the light particles would also be determined by speed of the source which ejected the light particles. The second theory was that light was essentially wave-like in nature and Space itself was a medium substance, called the luminiferous Aether or Ether, that both supported the sinusoidally changing light potentials and also determined their propagation velocity. It was taken that the Aether entirely filled the vacuum of Space such that light from the Sun, the stars and the galaxies was enabled to cross the enormous distances of Space to and from the furthest points of the Universe.

In many respects the postulated Aether medium is akin to a material medium which supports acoustic waves rather than the waves of light. The propagation velocity of acoustic waves is precisely determined by the pressure/density relationship of the particular material medium substance. Hence the acoustic propagation velocity differs radically from one material medium to another but nevertheless it is specific for a particular material. By analogy it was surmised that the Aether substance determined the propagation velocity of light waves.

If instead light were particles then the measure of the

speed of light emitted by a body would also depend upon the relative velocity of that body to the observer. Luminous celestial bodies frequently have a large radial velocity with respect to the Earth but the measure of the speed of their emitted light is always found to be constant. Consequently, for this reason - and particularly in order to explain the phenomenon of diffraction - the wave theory of light became favoured over the light particle theory and is now totally accepted.

However - and quite strangely - the theory of an Aether medium as the determinant of the speed of light is no longer accepted in Modern Physics despite the fact that modern physicists neither provide - nor feel inclined to provide - an alternative explanation for the determination of this universal constant. The reason for this paradox stems from the widespread acceptance of Einstein's Special Theory of Relativity - for in his theory Einstein effectively postulates that the Aether cannot exist.

It is true that the Aether theory of light propagation does indeed have an immediate difficulty. For if Space is a medium substance then it must be asked how large celestial bodies such as the Earth - made from solid matter substances such as rock and iron - move freely through the Aether substance with zero friction and zero turbulence. The answer to that question depends totally upon our understanding of the nature and construction of the fundamental matter particles (FMPs) of which all matter is constructed.

On the Structure of Fundamental Matter Particles

We intuitively believe matter to be a substance. And yet physicists are aware that atoms - the building blocks of matter - are predominately empty space, with the constituent particles, electrons, protons and neutrons, taking up an almost insignificant fraction of the total volume of the atom. In fact, atoms are very insubstantial objects. So the problem of atoms moving freely through the Aether is diminished somewhat by our knowledge of their construction. But nevertheless the problem has not been entirely removed. The initial difficulty with the Aether movement of atomic matter has merely been shifted to the movement of sub-atomic particles.

It naturally follows that if the Aether is a substance then mass particles cannot also be made of a substance, or energy would be continuously lost from the moving particle. A clue to the structure of matter may lie in Einstein's famous equation, $E = mc^2$. This equation states that matter can, under certain circumstances, convert into photons and vice versa. Thus it is possible that matter and energy are just different geometries of the electric wave (it is hypothesized in a later chapter that magnetism is not a fundamental force). Einstein's famous equation might therefore simply describe a change of geometry rather than a change of substance.

With the photon the electric wave moves in a straight line through Space at the speed of light. In matter the electric wave must move in some three dimensional rotatory manner such that the totality of the rotating wave group could remain stationary in the Aether as matter bodies theoretically may do. And yet every part of the stationary rotating electric wave group still propagates at the speed of light. In this particular model the maximum velocity of matter is naturally limited to the velocity of its constituent parts, the electric waves, which, as matter is observed to be limited to a velocity just below the speed of light, is indeed the case.

Both the straight line geometry and the rotatory geometry of the electric wave would need to be stable up to the point where circumstance would cause one geometry to flip into the other, thus translating mass into energy and vice-versa.

But there is a rather obvious further complication to the rotatory geometry postulate for matter particles. It is that matter particles are not merely stationary in the Aether. They may move at any velocity from zero to just below the speed of light. Thus the rotatory geometry of the mass particle must be complicated by some form of asymmetric 'screw' mechanism which enables the rotating electric waves to 'screw' there way through the Aether in a direction determined by the particular arrangement of the screw geometry. The Aether velocity of the mass particle is then a function of the magnitude of 'screw' carried by the particle geometry. According to this hypothesis, the Aether velocity of a mass particle is determined both in direction and in magnitude by the particular geometry of that particle.

It also follows from the observed acceleration of charged mass particles in force fields that the electric and gravitational fields must modify the 'screw' asymmetry as a function of the strength and the direction of the field. This change in 'screw' asymmetry might come about as a consequence of the ambient electric and gravity potential fields superpositioning upon the internal electric potentials of the mass particle. It is shown later that both the electric and the gravitational potential fields do indeed intrinsically superposition upon each other. Situated in a gradient of the local ambient field the electric potentials differ across the rotatory geometry of the mass particle. Thus the asymmetric geometry of the internal electric waves would be unbalanced and that unbalance might in turn modify the existing degree of 'screw' of the FMP. The degree of 'screw', and hence the velocity of the mass particle, would therefore be changed at a rate which is a function of the degree of the gradient of the ambient potential field.

It follows that the stability of the FMP 'screw' asymmetry, in the absence of accelerating fields, relates to the inertia of the particle.

The mass of a particle inversely describes its rate of acceleration in response to a unit gradient of accelerating field. This mass factor is not constan, for it is found to increase with the increased velocity of the particle, ($m = m_0/\sqrt{1 - v^2/c^2}$). Thus the greater the degree of 'screw' asymmetry of the particle the smaller the change in that asymmetry for an identical acceleration stimulous. This is intuitively just what we might expect from such a geometry.

This hypothesis of the internal structure of matter as a complex of electric waves removes the friction argument against the movement of matter through the Aether.

But there are still many other objections which have been raised against the Aether hypothesis. They are dealt with, one at a time, at various stages throughout this book.

Measuring the One-Way Speed of Light

It seemed that there was a relatively easy way of testing whether the Aether medium theory was correct or not. The Earth is unlikely to be stationary in the Aether as it moves around the Sun at an orbital velocity of 30km/s. The Sun also moves around the centre of the Milky Way and the Milky Way galaxy is also likely to be moving through the Aether. Therefore the total Aether velocity of the Earth, whatever it may be, adds to the propagation velocity of the Aether in one direction and subtracts from it in the opposite direction. It would appear that a measure of the one-way speed of light in many directions in the Heavens would eventually uncover this difference and thus directly give the Aether velocity of the Earth. A value for Earth's Aether velocity would then conclusively prove the existence of the Aether medium.

So how to go about measuring the speed of light. The obvious method to use is that employed for measuring the speed of any object eg. bullets, vehicles, etc. The moving object to be measured is made to pass along a track of a precise known length. A timing clock is started when the object passes the beginning of the track and is stopped when the object passes the end of the track. A division of the track distance by the elapsed time gives the average velocity of the object along the track.

There is one further important requirement which is that the entire measurement apparatus should not be accelerating, for an acceleration of the apparatus in the direction of the track means that the track length is effectively shortened or lengthened over the passage time of the moving object being measured. The apparatus must therefore be inertial. Theoretically speaking, the measurement apparatus exists within an inertial reference frame (IRF) of three distance dimensions and one time dimension. IRFs are a useful theoretical tool for physicists but they take on a particular importance in Special Relativity.

This simple apparatus needs to be modified somewhat when the measurement is of the speed of light. For a start, light pulse detectors need to be fitted at either end of the track which initiate a communication by some means to start and stop the timer when the object light pulse passes the detector.

But there is a fundamental problem with the measure of the speed of light which does not exist when measuring the

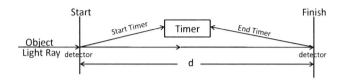

Figure 2.2: Speed of Light Measurement Apparatus 1

velocity of slower objects. Communication must be made over the distance from the light detectors to the timer. This then causes a time delay of significance relative to the measured time of the passage of the light ray down the track. In order to minimize the communication time delay the detector communication speed must therefore be made as high as possible. Indeed, it must be at the speed of light.

This can be simply arranged by certain changes in the apparatus. Instead of a light detector at the start of the track the incoming object light pulse is split into two parts by a beam-splitter situated at that point. One part of the light pulse continues down the track as the object light pulse and the other part goes to start the timer. Additionally, at the end of the track is placed a mirror which reflects the object light pulse back to stop the timer. Thus the light pulse to be measured also acts to start and to stop its own timer. This configuration, described in Fig (2.3), minimizes the start/stop communication time.

But now the question arises as to which is the best position to site the timer. Obviously it should be adjacent to the track as this minimizes the communication distances. Also, it seems to make sense to position the timer halfway down

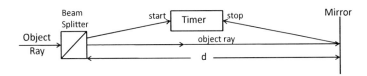

Figure 2.3: Speed of Light Measurement Apparatus 2

the track such that the start and stop timer communication times are equal. Then the two communication delay times cancel each other out.

But there still remains a serious problem of a fundamental nature with this solution. For the object light pulse now runs exactly parallel with the start the timer light pulse from the beam splitter as far as the timer. It follows that the two beams must always travel at exactly the same velocity for they travel in the same direction. They must therefore take exactly the same time to reach the position of the timer, irrespective of any value of the speed of light, and therefore irrespective of any value of Aether velocity of the apparatus. Hence this part of the apparatus, from the beamsplitter to the timer, can have no affect on the measurement whatsoever, and is entirely superfluous. The only possible position for the timer is now seen to be directly next to the beam splitter at the start of the track as described in Fig. (2.4).

But there arises a further problem. For it can now be seen that the modified apparatus is actually measuring the two way passage of a light beam - ie. out to the mirror and then back again. Thus the apparatus intended to measure a one way passage of light has now, of necessity, morphed

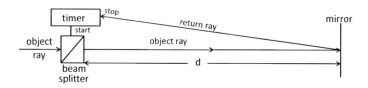

Figure 2.4: Speed of Light Measurement Apparatus 3

into something entirely different and contrary to our initial intentions.

It might, at first sight, be thought that this problem can be overcome by a variety of different solutions. For example, by the use of two identical timers - one at the start position and one at the end position of the racetrack. But the two timers need to be synchronized prior to the measure being taken. The act of synchronisation must be a communication by light ray between the two clocks over their separation distance. The time delay of the passage of the synchronisation pulse now unavoidably becomes a part of of the total measurement, albeit the synchronisation occured at an indeterminate time prior to the actual measurement. In fact, the synchronisation pulse equates to the start/stop timer communication pulse when only a single timer is used. Thus two separate timers are actually equivalent to a single timer positioned at one end of the track - which end merely depends upon the direction of the synchronisation pulse.

Alternatively, a clever person might suggest synchroniz-ing the two clocks when they are together at one end of the racetrack and then moving one of the clocks to the other end of the track prior to the measure, in order avoid the use of a synchronization pulse over the length of the track. This solution makes the assumption that the movement of a clock does not affect its reading. Now it is known that the velocity of movement causes the time dilation of clocks as a function of that velocity. But if the clock is transported very slowly from one end of the track to the other this effect will be minimal and may be disregarded. However there is still an implicit assumption that the slow transportation of a clock from one place to another does not affect its timekeeping by any other mechanism. It is demonstrated later that the slow transport of a clock creates exactly the same time effect as the passage of a light beam over the same distance.

The lesson to be learned from the above logic is that it is impossible to measure the one way speed of light. Only a two way measure can ever be made. Nevertheless several experiments have been made which purport to be able to measure the one way speed of light. In this the experimenters are mistaken.

Attempts to Measure the One Way Speed of Light
At first sight it would appear that Ole Roemer's method of measuring the speed of light is a one way method. Certainly the light from Jupiter's moon Io traverses Earth's orbit in a single direction only. Claims for a one way measurement all have one factor in common - instead of being stationary the timing clock is moved from one end of the measurement track to the other. In Ole Roemer's measurement the timing clock, stationed on Earth, moves from the nearest point to Io on Earth's orbit to the furthest point from Io. Thus the timing clock moves the diameter of the Earth orbit - the measurement track length - over the course of the measurement. Roemer's observation is an example of slow clock transportation.

It is not always obvious that a timing clock is moving in a measurement. For example a clock positioned on Earth is stationary with respect to Earth and hence to the human observer. At first sight the clock does not appear to be moving. But with respect to the Aether the clock is continuously moving in a complex pattern involving the Earth's rotation with Earth's orbital motion around the Sun.

The Two Way Axiom

A consideration of action and communication between bodies which stems from the lesson that a one way measure of the speed of light cannot be made, leads to a general axiom.

The Two Way axiom states:-
Meaningful communication between bodies must be two way

The movements of a two way communication may be made by any combination of velocities, ie. either by light or by matter.
Furthermore the out and the return movements may be separated by an indeterminate time interval.

For example, the Two Way axiom can be seen, acting in the operation of force fields. Consider two charges separated by a distance. Charge A accelerates charge B via its electric field. Simultaneously charge B accelerates charge A via its electric field. Thus A effects the postion of B and the changed position of B is communicated back to A via the electric field. The operation between charges is therefore two way at the speed of light, as the electric field is propagated at that velocity. An exactly similar operation occurs between massive bodies via their gravitational fields.

The requirement for the Two Way Axiom is also exampled in the reading of identical clocks separated by an un-known distance (see the chapter on Time).

The Two Way Axiom is important both in the under-standing of velocity effects and also in an understanding of the origin of the so called 'magnetic' effects. The effect of a two way communication is to effectively put the separation distance to zero despite the fact that the two matter bodies may still be physically separated. This point becomes more clear in the chapter on the Aether Theory of Velocity Effects.

The Two Way Measure of the Speed of Light

The measuring apparatus described in Fig (2.4) can be used to time the combined out and return passage of a light ray. The important question is whether the effect of the Earth's Aether velocity can still be detected with a two way rather than a one way measure.

We will assume for the sake of simplicity that the track of the apparatus happens to be directly inline with the Earth's Aether velocity. Thus the velocity of light relative to the apparatus is $(c + v)$ in one direction and $(c - v)$ in the op-posite direction, where the Aether velocity of the measuring apparatus is v. The time of flight of the light ray over the track distance d will be $d/(c + v)$ in the one direction and $d/(c - v)$ in the opposite direction.

The out and return time is therefore $2d/c(1 - v^2/c^2)$.

With the apparatus positioned transverse to Aether ve-locity the out and return time, determined from the geome-try of the situation (the full calculations are given in a later chapter), is:-

$$2d/c\sqrt{1 - v^2/c^2}$$

Thus there is a maximum difference in flight time between

the two orthogonal directions of

$$T_t/T_i = 1/\sqrt{1 - v^2/c^2} \qquad (2.1)$$

which is called the Lorenz factor, represented by the Greek letter γ.

So it seems that it is still possible to detect the Earth's Aether velocity, but now only to the second order effect. This requires a far more accurate apparatus. For example, if v/c was 1/10,000 (the Earth's orbital velocity) then v^2/c^2 will be just 1/100 million. To measure just one part in 100 million with sufficient accuracy requires a very stable and precise apparatus - particularly with respect to the timer.

The Timer

The difficulty for the timer can be seen if the track length is made to be, say, 10m. The out and return trip time is now only 67ns. A hundred millionth of that would be absolutely impossible to measure with any known clock of normal construction.

There is, however, one form of clock, called the Photon clock, capable of handling such short times. The time unit generator of the Photon clock consists of two mirrors facing each other separated by a set distance l determined by a matter rod.

A very brief light pulse is injected at one of the mirrors in the direction of the other mirror. The pulse travels to the opposite mirror where it reflects back to the first mirror and then reflects back again. Thus the pulse oscillates back and forth between the two mirrors with a period of $2l/c$. The light pulse is detected and counted every time it reflects from mirror 1. Of course, losses at reflection and at the detector mean that the pulse eventually diminishes to zero amplitude. However this does not matter in the manner in which the

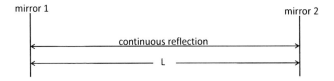

Figure 2.5: The Photon Clock

Photon clock is employed in the speed of light measurement apparatus as the Photon clock is not used in a continuous manner.

Normally, a time unit generator of any description creates a signal at the end/start of its generated time unit. These end/start signals are counted, stored digitally and then displayed as the elapsed time in digital form. Instead, when the Photon clock is employed as the timer in the speed of light measuring apparatus it is used in an analogue sense, and for only a single out and return pass of the light pulse.

The Photon clock is grafted to the measurement apparatus in the following manner. One end of the Photon clock is connected to the beam splitter such that the light pulse, previously used to start the timer, is injected into the Photon clock as its own light pulse. This light pulse travels to the mirror at the far end of the Photon clock and reflects back to the beam splitter.

Now if the length of the Photon clock is made identical to the race-track length then the two parts of the split light beam may be expected to return from the mirrors at the ends of each arm at the same instant. The beam-splitter, now acting in reverse, recombines the two returning pulses so that they interfere with each other to create a set of in-

terference fringes. These can be accurately observed with
a suitable optical apparatus. Any change in the journey
times of the two returning pulses relative to each other is
evidenced by a shift in the fringes. The number of fringes
shifted together with the frequency of the light source gives
the time difference between the two returning light pulses.

In the final apparatus the two identical arms are set at
90deg to each other such that the apparatus will experience
the expected time difference in the transit times of the two
arms, described by eqn.(2.1). This final apparatus is called
the Michelson-Morley (M-M) apparatus, shown in Fig.(2.6),
after its original inventors.

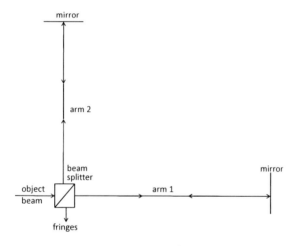

Figure 2.6: The Michelson-Morley Apparatus

In the M-M apparatus the photon clock timer - now just
one of the two identical arms of the apparatus - has totally
lost its significance as a timer. The conventional timing
of a one-way, or even a two-way, passage of a light beam

has become meaningless. The M-M apparatus merely races two light pulses along two identical length tracks pointing in orthogonal directions in order to measure any time difference between the two returning pulses.

The two way speed of light can, if desired, be measured using a clock of a more conventional type so long as the race track is made long enough such that the resolution of the clock is sufficient. Indeed the Fizeau experiment is exactly one example of such an apparatus. But although a measure of the velocity of light is obtained it really tells us nothing about the true speed of light. It merely tells us that our arbitrarily chosen time unit, the second, refers to a distance - a distance of 2.998×10^8 ms.

In order to fully understand the truth of that statement we need to examine the nature of Time itself.

Chapter 3

The Nature of Time

So what exactly is Time?

Modern Physicists call it a dimension - the fourth dimension - the other three dimensions being those of distance. They consider that Space has three distance dimensions which are inextricably linked with a Time dimension in a complex mathematical form called Space-Time. The concept of Space-Time was first invented by Hermann Minkowski in 1908 and was later adopted by Einstein as a basis of his General Theory of Relativity which gives an explanation of the operation of the force of gravity upon both light and matter.

But is Time really a dimension? For if it is then it is very different to the dimensions of distance. Bodies may move at greatly different rates (velocities) along a distance dimension, where-as time appears to move at a constant rate. Also, when moving at velocity in one direction through Space we can, at any time, reverse our movement and travel in the exact opposite direction. But in the Time dimension we can only move forward. Books and films on science fiction often fantasise about travelling back in time just as though

Time was a dimension similar to distance. But common experience tells us that travelling back in time cannot be done.

For example, if we moved back in time then by some way or another we could alter the circumstances of that prior time. In its turn, an altered past would develop into an altered present. That altered present might then disallow the very circumstances which brought about the backward time travel in the first place. Logic, together with experience, totally denies the possibility of time travel.

In order to understand time we must first study its effects. Simply put, time is experienced in change. And change appears to occur in two distinct forms. Firstly there is the change where an object, previously at one place, moves to a different place. This type of change is called movement and occurs at a great variety of velocities. Then there is a different form of change where the object may not move but instead undergoes a change in its size, structure or appearance. We can call this form of change 'structural change'.

Movement Change
The movement form of change is observed in the movement of the Sun across the sky every 24 hours, the passage of vehicles along a road, airplanes across the sky and ships over the oceans. Birds flying, fish swimming, animals and people walking and running are all examples of movement change. Snails crawling, Earth's tectonic plates creeping at a centimetre per year are further examples of movement change, but where the velocity is very low. Physics tells us that material bodies can move at any velocity from absolute zero up to just below the speed of light. The magnitude of the velocity is not material to the notion of movement change.

Now if time ever stood still we would agree that all this movement would stop. Every matter body would then equally have zero velocity. Yet when time moves forward at its constant pace moving bodies have an enormous variation in their rates of movement. Although movement is seen to be closely connected to the passage of time yet the variable rate of movement does not relate in any obvious way to the constant rate of the passage of time.

Structural Change

Structural change may be observed almost everywhere in one form or another. For instance it may be seen in the weathering of rocks, mountains and buildings. It may also be observed in chemical changes to materials such as in the rusting of iron. It can be seen in living plants and animals in their growth, aging and later, in their decay after death. Examples of structural change are endless.

Structural change seems to be entirely different to movement change and yet if it is examined closely it can be seen that there are common aspects. For instance the rate of structural change is slow to the point that the change only registers to the human brain if it is observed over very long periods. In some cases the required period may be hours and in others a noticeable change occurs only after hundreds, if not thousands, of years.

Also structural change, when closely examined, is seen to be effected by the movement of microscopically small particles, molecules or atoms, all of which are too small to see individually with the naked eye. For example, the weathering of rock occurs by the slow erosion of small grains of material from the surface of the rock. The rusting of iron results from the combining of oxygen atoms from the atmosphere with the iron atoms. The growth of plants and animals arises from the rearrangement of complex but minute molecules into the organised structure of the living thing.

A moments consideration of these thoughts leads to the conclusion that structural change is really movement change. But the movement is either too slow to be recognised by humans or it is of particles too small or too hidden to be seen by humans.

So we can finally state that the essence of time is movement, ie. a change in spatial position. The size of the moving body and the rate of movement are irrelevant to the concept.

But additionally we should also include the movement of photons as well as that of material bodies. Photons of energy move only at one velocity, the speed of light. Thus the evidence of time is inherent in the movement of both energy and matter, from one spatial point to another.

Although time is inherent in movement, yet movement cannot be the source of time, simply because the movement of matter can occur at an infinite variety of rates while time passes only at one rate. Movement is therefore a complex consequence of time and the real source of time still eludes us.

Clocks and Time Unit Generators

As time is characterised by movement then any constant rate of movement may be arbitrarily related to the constant rate of the passage of time. For example, if a body is made to travel at a constant velocity along a straight graduated course then an instant in time can be related to a distance marker on the course. Course marks may be positioned at equal separation so that they represent arbitrary time intervals - eg. one second, one hour, one day, etc. - at the particular velocity of the moving body. Clocks of this form have once been common. One particular example relied upon the steady drip of water from a vertical graduated container where the moving body was the level of the water. Another linear clock employed the steady burning of a long, graduated candle. Linear clocks have the practical disadvantage in that they have a life limited by the length of the candle or the height of the water container, ie by the limited length of a practical course - whatever form that course may take.

In constructing a stable and long-lasting time unit generator (a clock) it is far more convenient and practical to set a short fixed course and then cause the moving body to reverse direction at each end of the course. The back and forth movement can then continue indefinitely with each out and return journey creating a single time unit. Fractions of a time unit cannot be generated and so this form of clock is digital in nature, notwithstanding that the count of the time units may be displayed digitally or in an analogue fashion.

The reversal of a moving matter body at each end of the course cannot be undertaken instantly and so the velocity of the body is not constant over the course. In practice the velocity is usually constantly changing over the course. However if the total energy of the body in terms of the kinetic and the potential energy combined is maintained constant then the time unit period will also be constant. Examples of this form of time unit generator are the pendulum, the balance wheel and the crystal oscillator. Almost all clocks and watches employ one of these three forms of time unit generator.

An implementation of the oscillating body time unit generator which is theoretically the most perfect is the photon clock, shown in Fig.(2.5) and described in the previous chapter. By employing photons as the moving body of the clock the velocity is constant - and furthermore universally constant (in Euclidean Space). Every photon time unit generator will therefore create identical time units if the separation distance of the two mirrors is identical.

Time and the Structure of Matter

Although clocks enable us to measure the passage of time we have advanced no further in our understanding of time itself - or the cause of it. A matter body moving at a constant velocity enables us to measure the passage of time along a marked course but as the velocity may be from near zero up to the speed of light it would appear that the source of universal time cannot lie in any one body; and therefore must reside in none of them. However there is a clue towards an answer. The variability of velocity only applies to matter. The velocity of photons and the propagation velocity of the electric and the gravitational potentials is a universal constant. Therefore if matter did not exist time would be analogous to distance travelled at the speed of light.

Now the Hypothesis of the Structure of Fundamental Mass Particles dictates that matter is constructed of a rotatory geometry of electric waves. Thus, although the totality of the rotating wave geometry moves through the Aether at a variable velocity the constituents of a particle, the electric potentials, all move (although not necessarily in straight line paths) at the speed of light. The consequence of the FMP Hypothesis is that the Universe, in both matter and energy, consists solely of electric and gravitational potentials - all moving at the same velocity.

Thus Time is now analogous to distance travelled at the speed of light. A separate Time dimension just does not exist.

This is not to say that time does not exist. It certainly does exist, just not as a dimension. Instead, Time is embedded within the Aether and is evidenced in the local propagation velocity of the Aether. Time is therefore local rather than Universal.

Time is not Universal Constant

In a Universe where the velocity of every photon and every matter particle is determined as a certain fraction of the speed of light it is illuminating to consider what effect a change in the rate of the passage of time would have upon the Universe. For example, if the speed of light was changed by a factor of ten then what affect would that have upon the working of the Universe. The answer is simply none whatsoever. The Universe would carry on just as before. Every matter velocity would still remain exactly the same fraction of the new speed of light.

What matters to the evolution of the Universe is the rate of movement of bodies relative to the speed of light and relative to each other. In other words it is the degree of 'screw' geometry of individual mass particles which is important as this determines the fraction of the speed of light. The universal rate of passage of time is of no importance to the Universe but if the rate of passage of time varied locally from place to place then there would indeed be an effect. Local changes of time in the form of local differences in the speed of light are indeed observed as a consequence of the presence of mass. Gradients of the speed of light are associated with gravitational acceleration. This aspect of time is examined in the chapter on gravity where an Aether theory of gravity is described.

If the rate of passage of time seems to be constant throughout the Universe this is simply because the speed of light is fairly constant in the areas of Space which we examine compared to our own local speed of light. A difference in time scale between spatial points can be seen in the red or blueshift of the frequency emitted from known distant sources at various points in the Heavens.

When discussing non-gravitational matters such as velocity effects it is assumed for the sake of simplicity that the Aether is Euclidean (zero mass). Then the passage of time is everywhere the same and the speed of light is everywhere a constant.

The Artificial Time Dimension

As it is the case that a time dimension does not exist in reality then what is it that we arbitrarily and artificially call the time dimension. The concept that time is distance divided by the speed of light (d/c) is equally applicable to distances measured within inertial reference frames (IRFs) as those measured through the Aether. The Two-Way Axiom dictates that one-way measures are not possible so the distance analogous to time in an IRF is an out and return distance. This book employs this artificial concept of a time dimension throughout because readers are familiar with and expect the use of a time dimension.

The two-way distance travelled by a light ray between two spatial points within an IRF will be shown to be longer when that IRF is moving through the Aether than if it were stationary in the Aether. If the two points in question are the two ends of a photon clock then it can be seen that the time unit of the clock is increased. This effect is called time dilation.

The transformations of time and distance between the Aether and a moving IRF, and also between IRFs, are described and developed in later chapters.

Matter Time

The photon clock, in which light reflects back and forth between two mirrors placed a known separation distance apart, employs the artificial time dimension of d/c directly, where d is twice the separation of the photon clock mirrors. However it is not immediately obvious how a pendulum clock or balance wheel watch derive their time units in keeping with the artificial time dimension. But it has been shown that these mechanisms rely upon a constant average velocity between two points separated by a fixed distance, albeit that the instantaneous velocity may vary between the two points. Consequently the average velocity of the moving matter mechanism is fixed as some fraction of the speed of light via the degree of screw deformation within the mass particles of the clock mechanism. Thus diverse mechanical time unit generators are nevertheless closely related to the principle of the photon clock although, at first sight, there seems little connection.

Time unit generators mark off the passage of time in arbitrary sized equal amounts, eg seconds, in order that these units can be easily counted and stored as digital quanta. However the shortest time units empoyed by humans, such as pico-seconds, are enormously greater than the physical time units of the Aether. For all practical intents and purposes physical time, as evidenced in the propagation velocity of the Aether, may be considered to advance in an analogue manner.

The Reading of Distant Clocks
The reading of a distant clock B by an observer must be made in accord with the Two Way Axiom by means of an out and return light pulse or equivalent.
The one way transit time delays all communications both to and from a distant clock but the delay cannot be directly measured. Instead the one way transit time must be calcu-

lated as one half of the two way transit time - which can be measured. On the assumption that the observer possesses an Aether velocity the calculated delay time is not correct in absolute terms but nevertheless it is true for the observer. The explanation of this is developed over later chapters. The calculation of the one way transit time for each and every reading automatically allows for any change in separation distance since the last reading.

Simultaneity

The simultaneity of separated events is an arbitrary decision by the observer. He may consider that if he observes two events at the same time that they are simultaneous. On the other hand he may calculate the one way transit times from the distant sources of the events and determine simultaneity after an allowance for the measured transit time. Therefore there is no absolute defintion for simultaneity.

Time is a Concept only for Living Creatures

Consider two particles, A and B, moving in different directions at velocities $0.1c$ and $0.2c$ respectively, but which for an instant occupy the same spatial point, O. When particle A has moved distance d from O then particle B has moved distance $2d$ from O. We know this because we can make calculations from our knowledge of the two velocities over the same time period. By assuming the speed of light to be constant over the two paths we have made the passage of time to be constant. Even when the speed of light varies over the Space in question we can still make our calculation so long as we know exactly how c varies through that Space. But the calculation has become much harder and the amount of knowledge required is considerably more.

Although we humans may calculate these positions neither body A nor body B knows, nor has the ability to calculate, the position of the other body once they have left point O. If one body tries to discover the position of the other by means of communication using a light ray the result will be incorrect. This is because the other body has moved on during the transit time of the communication.

It is the case that inanimate bodies moving through the Aether do not need to know, and therefore do not need to predict, the position of other bodies, for they have no reason to concern themselves with their own fate. Only living creatures need to make predictions and consider the future. This is because living creatures have a need to pro-create and therefore they cannot leave their lives to chance. Their lives have to be organised around the complex process of reproduction for which certain future resources and actions are required. Even unthinking plants must consider the future effect of the seasons and climate, although their consideration takes place within their genes as a consequence of evolution rather than within a thinking brain.

The future is important to living things. Consequently living things need to invent a time dimension.

Chapter 4

The Michelson-Morley Experiment

The final version of the apparatus for the measurement of the speed of light shown in Fig.(2.6) is called the Michelson-Morley apparatus after the two experimenters, Albert Michelson and Edward Morley, who first employed it in the year 1887 at the Case Western Reserve University. Their object was to measure the Aether velocity of the Earth through a differential measure of the speed of light over orthogonal directions. Primarily for his work on this apparatus Edward Michelson was awarded the Nobel Prize in Physics in 1907. The out and return transit time calculations for the two arm of the apparatus have been previously given but are repeated here in more detail.

According to the Hypothesis on the Structure of Fundamental Mass Particles described in the first chapter material bodies, such as the M-M apparatus, pass through the Aether with the same facility as a light ray. Thus the M-M apparatus is totally transparent to the Aether 'wind'.

The Transverse Arm

An arm of the Michelson-Morley (M-M) apparatus is transverse to the Aether velocity of the apparatus. Fig.(4.1) is

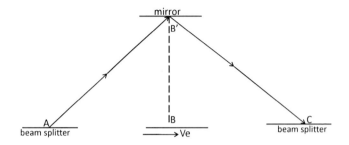

Figure 4.1: The Transverse Arm

a representation of the path of the light ray through the Aether.

The light beam initially leaves the beam-splitter at Space point A and travels towards the mirror at the far end of the arm. By the time the beam reaches the mirror, the mirror has moved on at the Aether velocity, V_e, of the apparatus to Space point B'. In that same time the beam-splitter has moved on to Space point B directly opposite B'. After reflection at the mirror the beam travels back to the beam-splitter. But over the return time of the light beam the beam-splitter has moved on to Space point C.

Consider the triangle ABB'. The light beam travels side AB', the length of which is proportional to the speed of light through the Aether. In the same time the beam-splitter travels side AB, the length of which is proportional to the Aether velocity of the apparatus, V_e. Side BB' is the length of the apparatus arm, L.

By Pythagoras the length of the diagonals AB' and $B'C$

calculate to be $L/\sqrt{1 - V_e^2/c^2}$ or $L\gamma_e$.

The time t_t taken to traverse AB and BC from the beam-splitter to the mirror and back again is therefore:-

$$t_t = 2L\gamma_e/c$$

The journey time of the light beam when $V_e = 0$ is $2L/c$, called t_0. Thus:-

$$t_t = t_0\gamma_e$$

In Fig.(4.1) the light beam appears to be 'launched' into the Aether at an angle in order to strike the mirror at point B. It only appears this way when viewed from the Aether. When viewed from the IRF of the apparatus the light beam source is aimed directly at the mirror and the reflected beam returns directly back to the source.

The In-line Arm
An arm of the M-M apparatus is in line with the Aether velocity of the apparatus.

Figure 4.2: The In-Line Arm

The light beam leaves the beam splitter at Space point A and travels to the mirror which was initially at point B but in the meantime has moved on to Space point B' as a result of its Aether velocity, V_e. The light pulse reflects from the mirror back to the beam-splitter which, in the meantime, has moved on to Space point C.

The outward time, t_o, is given by:-

$$t_o = (L + t_o V_e)/c$$

Therefore

$$t_o = L/(c - V_e)$$

It follows that the return time, t_r, is given by:-

$$t_r = L/(c + V_e)$$

hence

$$t_o + t_r = t_i = 2L\gamma_e^2/c$$

where t_i is the out and return time for the inline arm. Substituting for $t_0 = 2L/c$ gives:-

$$t_i = t_0 \gamma_e^2$$

The ratio of the inline and the transverse times is therefore:-

$$t_i/t_t = \gamma_e.$$

It can be seen that the travel time for the arm parallel to the Aether velocity is longer than that of the transverse arm by the factor γ_e. Therefore a measure of the maximum ratio of the transit times of the two arms of the M-M apparatus - obtained by taking measurements of the transit times over all directions in Space - would be expected to determine the factor γ_e and hence the magnitude of the Earth's Aether velocity.

The truth or otherwise of the Aether hypothesis would then be apparent.

The Accuracy Required

The velocity V_e of the Earth through the Aether is a combination of the Earth's orbital velocity around the Sun, the Sun's orbital velocity around the centre of the Milky Way and the velocity of the galaxy itself through the Aether. Due to the lack of information in 1887 it was possible that our galaxy was neither rotating nor moving through the Aether, so the apparatus needed to be capable of measuring the effect of the Earth's orbital velocity of $30km/s$ alone. Therefore the maximum value of the factor V_e/c might be just 10^{-4} and the value of $(V_e/c)^2$ just 10^{-8}.

The M-M apparatus needed to be capable of measuring this very small fraction to an acceptable degree of accuracy.

The Construction of the MM Apparatus

Michelson and Morley spent a considerable amount of time and money creating a version of their apparatus with more than enough accuracy to detect the Earth's Aether velocity. In their apparatus, the light was repeatedly reflected back and forth along the arms, increasing the path length to 11m. At this length, the expected Aether velocity would generate a shift of about 0.4 interference fringes.

In order to remove spurious effects the apparatus was located in a closed room in the basement of a stone building which eliminated most thermal and vibrational effects. Vibrations were further reduced by building the apparatus on top of a huge block of marble floating on a pool of mercury. The mercury pool allowed the device to be turned, so that it could be rotated through the entire range of possible angles to the Aether 'wind'. The experimenters calculated that movements of about 1/100th of a fringe would be detectable.

Even over a short period of time some sort of effect would be noticed simply by rotating the device, such that one arm alternately rotated into the direction of the Aether 'wind' and then across the 'wind'. Over longer periods, such as day/night or yearly cycles, the apparatus would point in every possible direction with respect to the Aether wind such that the maximum value of the difference in the transit times of the two arms must eventually be discovered.

During each full rotation, with the apparatus in its most advantageous position relative to the Aether 'wind', each arm would be parallel to the Aether 'wind' twice and transverse to the Aether 'wind' twice. This effect would show interference fringe readings in a sine wave formation with two peaks and two troughs. Additionally, if the Aether wind arose only from Earth's orbit around the sun, the wind would fully change directions east/west during a 12 hour period.

The Result

Within the limits of the accuracy of their apparatus, Michelson and Morley finally accepted that they could not detect a fringe shift which could be attributed to the movement of the apparatus with respect to the Aether 'wind'. Thus the apparatus gave a null result for Earth's Aether velocity.

This result was considered to be utterly remarkable as the hypothesis of the Aether was widely accepted throughout the physics community at the time. Hence doubt was immediately placed on the accuracy of Michelson and Morley's conclusion. In order to remove this doubt the experiment was repeated with various refinements to the apparatus and by a variety of different experimenters over both the 19th and the 20th century.

In recent times different versions of the Michelson-Morley experiment have become commonplace. Lasers and masers amplify light by repeatedly bouncing it back and forth inside a carefully tuned cavity, thereby inducing high-energy atoms in the cavity to give off more light. The result is an effective path length of kilometers. Better yet, the light emitted in one cavity can be used to start the same cascade in another set at right angles, thereby creating an interferometer of extreme accuracy.

The first such experiment was led by Charles H. Townes, one of the co-creators of the first maser. Their 1958 experiment put an upper limit on drift, including any possible experimental errors, of only 30 m/s. In 1974 a repeat with accurate lasers in the triangular Trimmer experiment reduced this to 0.025 m/s, and included tests of entrainment by placing one leg in glass. In 1979 the Brillet-Hall experiment put an upper limit of 30 m/s for any one direction, but reduced this to only 0.000001 m/s for a two-direction case. A year long repeat known as Hils and Hall, published in 1990, reduced the limit of anisotropy to 2×10^{-13}.

The final conclusion has to be that the transit times of the two arms of the Michelson-Morley apparatus are exactly equal for all directions in Space.

Explanations for the Null result
Various explanations, some quite implausible, were immediately put forward to explain the null observation.

One explanation was that a layer of Aether remained attached to the Earth, by some means or other, as it moved through the main body of the Aether. This effect was called 'Aether dragging' or 'entrainment'. It implied that the local Aether surrounding the Michelson-Morley apparatus moved with the apparatus rather than with the main body of the Aether further out in Space. A number of experiments were carried out to investigate the concept of Aether dragging. The most convincing was carried out by Hammar (1935), who placed one arm of the interferometer between two huge lead blocks. If the Aether were dragged by mass, the blocks would, it was theorised, have been enough to cause a visible effect. Once again, no effect was seen.

Walter Ritz's emitter theory (the ballistic theory) of photon production, was consistent with the results of the experiment in not requiring an Aether. Furthermore his theory was more intuitive and paradox-free. However it also led to several optical effects that were not seen in astronomical photographs, notably in observations of binary stars in which the light from the two stars could be measured in an interferometer. If the Ritz's postulate was correct, the light from the stars should cause fringe shifting due to the velocity of the stars being added to the speed of the light - but no such effect could be seen.

The Sagnac experiment placed a modified M-M apparatus on a constantly rotating turntable; the main modification being that the light trajectory encloses an area. In doing so any ballistic theories such as Ritz's could be tested directly, as the light going one way around the device would have a different length to travel than light going the other way (the eyepiece and mirrors would be moving toward/away from the light). In Ritz's theory there would be no shift, because the net velocity between the light source and detector was zero being both mounted on the turntable. However in this case an effect was seen, thereby eliminating any simple ballistic theory. This fringe-shift effect is used today in laser gyroscopes.

The need for these various explanations, which now seem rather bizarre, resulted directly from the concept of the Aether as a substance at a time when matter itself was also considered to be a substance.

Initially, some explanations chose to modify the concept of the luminiferous Aether by attempting to give it liquid or even gaseous qualities, such that points in the Aether were capable of movement with respect to other points. But if points in the Aether were capable of moving with respect to each other beyond the confines of a matrix then the concept of a constant velocity of light through the Aether is effectively destroyed.

The Fitzgerald-Lorenz Contraction Effect

Eventually a more logical explanation of the null effect was introduced based upon the concept that matter was not as substantial as previously believed. The idea was suggested by Irish physicist G. F. Fitzgerald and independently by the Dutch physicist H. A. Lorentz. They postulated that the length of the arm of the apparatus in line with Aether velocity was contracted, as a consequence of that velocity, by the factor $1/\gamma$, while the transverse arm was unaffected. This degree of contraction ensured that the light pulses from the two arms returned at precisely the same time so as to give the null effect observed. This is called the Fitzgerald-Lorentz contraction effect (FLCE).

If neither arm is fully inline with the Aether velocity but an angle to it then each arm can be considered to have a component inline with Aether velocity and a component parallel to it. The addition of these component parts makes two complete theoretical arms, one inline and the other parallel. Thus the two transit times are equal for any possible orientation of the apparatus.

To be exact the FLCE does not need to be an absolute contraction of the in-line arm. It only needs to be a contraction relative to the transverse arm. As there are two dimensions - call them y and z - to the transverse plane then these dimensions are taken to be the standard against which differences in the length of the inline arm, the x dimension, are related.

It should be realized that Fitzgerald and Lorenz did not suggest that the contraction effect applied solely to the arms of the Michelson-Morley apparatus. Their proposal was that the FLCE applied equally to all matter, whether that matter was rock or rubber, solid or liquid. Thus all the matter in the Universe is length contracted in the direction of its particular Aether velocity and as a function of that velocity. Not only is an arm of the Michelson-Morley apparatus contracted but so is the human observer and all of his measuring instruments, the laboratory which houses the apparatus, and the whole Earth upon which the laboratory sits.

No matter how high the Aether velocity and how large the contraction effect it cannot be detected by any form of measurement by an observer co-moving with the body. This is simply because his measuring apparatus, eg. a measuring rod, is equally contracted to the same extent as the distance or length being measured.

The contraction effect is very small for even quite fast velocities. At a velocity of 1000km/s the contraction is only 6 parts in a million. The effect only becomes significant at velocities approaching the speed of light.

It should be realized that the FLCE was a very revolutionary concept in 1887. At that time matter was considered to be a substantial solid which required considerable force to accomplish even a very small compression. Where-as, according to the FLCE, at near the speed of light and in the absence of any forces, matter is contracted to almost zero dimension in the direction of movement. Matter had suddenly become insubstantial. It was many years after the Michelson-Morley experiment before the realization dawned that atoms, the building blocks of matter, consisted almost entirely of empty Space.

The Kennedy-Thorndike Experiment

A modified version of the Michelson-Morley experiment was carried out by the experimenters R. J. Kennedy and E. M. Thorndike in 1932 some 45 years after theoriginal M-M experiment and 27 years after Einstein's seminal paper, the Special Theory of Relativity. Their novel approach was to make one arm of the Michelson-Morley apparatus very much smaller than the other arm.

The experimenters stated;-

'The principle on which this experiment is based is the simple proposition that if a beam of homogeneous light is split [...] into two beams which after traversing paths of different lengths are brought together again, then the relative phases [...] will depend [.] on the velocity of the apparatus unless the frequency of the light depends [.] on the velocity in the way required by relativity.'

Thus if there was an Aether 'wind' the new apparatus might give a non-null result despite the Fitzgerald-Lorenz contraction effect. The Kennedy-Thorndike experiment indeed still did give a null result and it was therefore generally concluded that there really was no Aether 'wind' and hence no Aether. However it is necessary to closely examine the final part of the experimenters' statement - *'unless the frequency of the light depends on the velocity in the way required by relativity'* - in order to fully understand the reasoning upon which a conclusion can be based.

Einstein's Special Theory of Relativity predicts the phenomenon of time dilation, which occurs as a function of relative velocity. Thus, according to Relativity the frequency of a light source suffers time dilation, ie lowers frequency, when moving relative to an observer. But as the light source of the K-T apparatus is not moving relative to the observer no time dilation can be expected from Einstein's prediction. However, Kennedy and Thorndike suggested that a similar time dilation effect to that of Einstein's theory might possibly occur but necessarily as a function of Aether velocity - in which case a null effect might still be observed.

So does time dilation occur as a part of Aether theory?

It should be remembered that each arm of the M-M apparatus is identical to a photon clock. The transit time for a light ray to pass back and forth along the transverse arm is given by $t = t_0\gamma$. Thus the length of the time unit of the transverse photon clock increases/dilates with Aether velocity. Exactly the same effect occurs in the inline arm after the FLCE contraction of that arm is taken into account.

It is a simple step to realise that if time dilation is experienced in one form of time unit generator, ie. the photon clock, then it must also occur in all forms of clocks, and hence in matter generally. Thus time dilation as a function of Aether velocity also occurs in the light source of the M-M apparatus, such that the wave-length of that source is affected. This shift of wave-length of the light source accounts for the K-T null result.

It is the case that Aether theory predicts both length contraction and time dilation. This is simply because time and distance are analogous to each other, as previously explained in the chapter on time.

Several other experiments have been undertaken which purport to deny the Aether and the FLCE explanation of the Michelson-Morley experiment. The Trouton-Noble and the Trouton-Rankine experiments are two examples. The statements that these experiments and others prove that the

Aether does not exist all suffer from the same weakness. In calculating whether the obtained null result is a prediction of Aether Theory the experimenters must first understand that theory and apply it correctly. In particular the experimenter may not realise that time dilation is an integral part of the Aether theory of velocity effects, just as Kennedy and Thorndike did not.

Furthermore, errors in the calculations are very likely when the person making these calculations does not fully understand the Aether theory that he is attempting to employ. A single error in his calculations gives a non-null result which then 'disproves' the existance of the Aether. Furthermore, the average physicist has been weaned on Special Relativity and has an inbuilt bias towards that theory. Given the task of calculating whether an experiment proves or denies the Aether he is unlikely, subconsciously, to want to disprove his own beliefs.

On the other hand Special Relativity does not need to prove how a null result comes about for the theory simply postulates that the observer can never obtain a non-null result in his own IRF.

Is There a Limit to Observation

It can be seen that the initial objections to the FLCE explanation of the M-M null result rested upon a false understanding of the structure of matter and upon an equally false understanding of Time. Nevertheless Aether theory still had further objections to overcome.

A common problem that many people have with Aether theory is that the Aether velocity of a body cannot be detected by any local experiment whatever. Take magnetism for example. If a magnet is passed through a loop of wire a voltage is induced across the ends of the loop. The voltage is a function of the relative velocity of the wire to the magnet. The induced voltage is the same if either the wire or the magnet is moved. The absolute velocities of the wire and magnet do not appear to have any relevance. In fact the whole of the laws of physics, not just those of electromagnetism, are observed to be identical within all IRFs, irrespective of the relative velocities of those IRFs to any other body.

If one cannot detect the effect of Aether velocity in any experiment whatever then it is understandable that the Aether is taken to be non-existent. We like to believe that we have the power with modern science and modern instruments to detect anything and everything, but the false assumption is made that our instruments are not modified by their Aether velocity.

Our Aether Velocity Determined

In fact the Aether velocity of the Earth can be detected, although not by local observation. In 1964 two radio astronomers, Amo Penzias and Robert Wilson, detected microwave emissions emanating from Space. The frequency of these emissions covered a band which peaked at a wavelength of 1.9mms. The unusual aspect of these microwave emissions is that they did not come from any particular celestial body but instead came equally from every direction in Space. They seemed to arise from Space itself. These Space emissions are now called the Cosmic Microwave Background Radiation, or CMBR for short.

The emission band was identical for all directions except that the band was shifted to a higher frequency in one particular direction and was equally lower in frequency in the opposite direction. It seemed that this frequency shift could most easily be explained by the Doppler effect caused by the velocity of the Earth relative to the source of the CMBR.

The degree of Doppler shift indicates that the Earth is moving at a velocity of 360 km/s with respect to the source of the CMBR in the direction of galactic longitude, l = 276deg., b = 30 deg..

Thus there exists a Universal Source of radiation through which Earth is moving. That Source must originally have been material bodies such as stars (although believers in the Big Bang may not accept that conclusion). These bodies would have been moving through the Aether, but in random directions such that their average Aether velocity was zero. So on average, these source bodies marked the position of the Aether. Consequently the CMBR is effectively sourced from the Aether and the CMBR dipole anisotropy of 360km/s is the true velocity of the Earth through the Aether.

The M-M experiment may have failed to detect the Earth's velocity through the Aether but the CMBR measurement suceeded.

However a full and proper understanding of Aether theory was not available at the time of the M-M experiment and neither had the CMBR radiation been detected. Consequently alternative explanations of the M-M null result were still pursued.

Some 18 years later a revolutionary explanation of the null result was offered by an unknown 26 year old patent clerk named Albert Einstein.

Chapter 5

Special Relativity

The matter contraction explanation proposed by Fitzgerald and Lorenz was one solution to the unexpected null result of the M-M experiment. But the FLCE required that solid matter must contract to almost zero dimensions at near luminal velocity without the application of any compressive force whatever. At the time, in the late 19th century, this was not an easy idea to accept. At first sight Albert Einstein's alternative explanation appeared to be very simple and straightforward. He merely suggested that if the speed of light was measured to be constant in all directions then that was because it actually was constant in all directions. Although Einstein's solution seems simple it will be shown to lead to horrendous complications.

As the Earth-based M-M apparatus could not in anyway be special within the enormity of the whole Universe then it must follow that the speed of light is constant and isotropic for every observer. Also there must be an infinity of potential observers within the Universe. More precisely, Special Relativity refers to observations between theoretical IRFs rather than between observers so it also follows that there must be an infinity of theoretical IRFs within the Universe - and there may be considerable relative velocity between them.

The Postulate of the Special Theory

Einstein's postulate for his Special Theory of Relativity (SR) stated that:-

The laws of physics are equivalent for all inertial frames of reference

A secondary postulate, derived from the main postulate, states:-

Light propagates in vacuo in all directions at all times and in all IRFs at the same velocity.

The first postulate is known as the Principle of Special Relativity and the second is called the Light Principle.

In his theory Einstein gives great significance to inertial reference frames which it can be seen are incorporated within both of his postulates. In doing so he elevates what was previously merely a theoretical mathematical tool into a quasi-physical entity.

Einstein determined that distance and time within an IRF is measured with reference to a standard length unit, eg. one meter, in the form of a material rod, and a standard time unit, eg. one second, determined by some form of material clock. These materially constructed units must be stationary within the IRF in question.

IRFs cannot themselves be directly detected. Only material bodies stationary within the IRF and their associated electric and gravitational fields can be detected.

It can readily be seen that both of Einstein's postulates deny the existence of an Aether, as the laws of physics must naturally be different in the Aether compared to those within a theoretical IRF moving relative to the physical Aether. For example, if light is isotropic in the Aether then it cannot be isotropic in an IRF moving through the Aether. This is contrary to the Principle of Relativity and to Einstein's Light Principle. Consequently, Einstein's postulates deny the concept of absolute motion for matter bodies as, in the absence of the Aether, there no longer exists an absolute reference frame relative to which velocity may be measured. Instead, Einstein argued that the motion of a body can only be determined relative to another body or its IRF. Therefore in Special Relativity, velocity is an arbitrary parameter as its magnitude depends entirely upon the choice of reference body.

The Principle of Special Relativity is given in greater detail as:-

A frame in uniform translatory motion with respect to an inertial reference frame cannot be distinguished from that inertial frame by any physical experiment whatever.

At first sight the Principle of Special Relativity appears to be quite a profound and important statement in physics. However it is, in effect, merely stating that all IRFs are indistinguishable from each other when examined by experiment. In reality this principle should be rather more precisely defined by the substitution of the phrase 'local experiment' instead of just 'experiment. The word 'local' denies the employment of astronomical observations where-by, it is maintained, that certain differences between IRFs are indeed observed.

All local experiments to date which have been designed to test this principle - and there have been a great many - agree with the Principle of Special Relativity and to a remarkable degree of accuracy. However, it must immediately be said that Aether theory also entirely accepts the Principle of Special Relativity - hence the Principle really ought to be renamed.

The Principle of Relativity and Einstein's first postulate are actually saying rather different things without, it might seem, the intention to do so. The Principle states that no two IRFs *can be distinguished* from each other while the postulate states that the laws of physics *are* equally valid in all IRFs. The difference between these two statements is profound. It is the difference between fact and observation. For example, the Light Principle states that the speed of light *is* constant and isotropic in every IRF. If instead Einstein meant that light is just measured to be constant and isotropic then Einstein would be saying nothing different to Aether Theory.

The Predictions of the Special Theory

The Special Theory of Relativity (SR) makes four major predictions. These are the distance contraction and time dilation of IRFs moving relative to an observer as a function of that relative velocity, and the associated length contraction, time dilation and mass increase of matter bodies as a consequence of being stationary within that IRF.

Finally, Einstein derived the mass/energy translation formula, $E = mc^2$. Thus we have:-

$$l = l_0/\gamma_r, \quad t = t_0\gamma_r \quad \text{and } m = m_0\gamma_r$$

l_0, t_0 and m_0 are the values of length, time unit and mass when the relative velocity is zero. $\gamma_r = 1/\sqrt{1 - V_r^2/c^2}$ where

V_r is the relative velocity.

The major predictions of SR are length contraction and time dilation as the other predictions of velocity effects may be derived from these two.

According to SR, the M-M apparatus is not length contracted from the experimenter's point of view as the apparatus has zero relative velocity with respect to the experimenter. On the other hand an observer passing by in a rocket ship at some relative velocity would observe the M-M apparatus to be contracted in the direction of his relative velocity. Additionally, if a clock was situated by the side of the M-M apparatus and the observer in the rocket ship also had an identical clock with him then the rocket ship observer would note that the M-M apparatus clock ran slow relative to his own clock.

The Lorenz Transforms

As a step on the way to the generation of the final predictions of Special Relativity, Einstein first calculated the well known Lorenz equations for transforming dimensions in one IRF to those of a second IRF moving at velocity relative to the first. These equations were originally calculated, although not in the final accepted form, by an almost equally famous scientist of the period named H. A. Lorenz. Einstein constructed the Lorenz Transforms by employing a scenario based upon the postulates of Special Relativity. Einstein's construction and his calculations are described in many, many text books on Relativity, so there is no need to describe them here.

The Lorenz Transform equations are;-

$$\xi = \gamma(x - vt) \tag{5.1}$$
$$\eta = y \tag{5.2}$$
$$\zeta = z \tag{5.3}$$
$$\tau = \gamma(t - xv/c^2) \tag{5.4}$$

Where ξ, η, ζ and τ are the dimensions of IRF S and x, y, z and t are the dimensions of IRF S' where the co-ordinates x and ξ are in line with each other. The relative velocity v between frames S' and S lies in the x, ξ direction.

The Lorenz Transforms are particularly useful when they are used to generate a further equation which predicts the velocity of a moving body observed from two different IRFs moving relative to each other. The calculations for this Velocity Transform equation are given in any standard text book on Relativity but, due to its importance, the full derivation is given in Appendix 4.

The Lorenz velocity transformation equation is:-

$$\gamma_u/\gamma'_u = \gamma_v(1 + u'_x v/c^2) \tag{5.5}$$

γ_u, γ'_u and γ_v are the Lorenz functions involving velocities u, u' and v where u is the velocity of the body when observed in IRF S and u' its velocity when observed in IRF S'. The relative velocity between the two IRFs is v.

Is Special Relativity a Successful Theory
The answer to that question is that it is a qualified and a partial success.
Certainly it is widely accepted that SR correctly predicts the

observations so far made of length contraction, time dilation
and mass increase with respect to many varied phenomena
and also, in some cases, to a remarkable degree of accuracy.

However, SR is unable to correctly predict the result of
one particular type of experiment, called the Twins Paradox.
The essence of the Twins Paradox type of experiment is that
the observer changes IRF within and during the experiment.
The response to this failure has been that modern physicists
have either ignored the problem or instead pretended that
it did not exist. The Twins Paradox is such an important
experiment that this book devotes a whole chapter to it.

Reciprocal Observations

The essence of the logical difficulty of Einstein's theory is
exposed when two bodies moving at a relative velocity make
observations of each other. We may take the two bodies to
be identical and accurate clocks called A and B. Clock B
moves through the IRF of clock A at velocity v which also
means that clock A moves through the IRF of clock B at
the same velocity. It should be noted that the velocity v is
relative to the IRFs of the bodies rather than between the
actual bodies.

Each clock observes the other over the separation distance (the magnitude is not important to the observations)
by means of a two way communication by light ray and making the necessary calculation to allow for any change in the
separation distance. Each clock reads the time held on the
other clock at the start and the end of a chosen time period
determined by his own clock. The length of the observed
time period is simply obtained by subtracting the first reading from the second reading. Thus the observer clock can
compare the time period measured by the observed clock
with his own determination of that period.

According to Special Relativity each clock acting as the observer finds that the other clock measures a shorter period than his own clock to exactly the same degree. Thus clock *A* observes *B* to run relatively slowly while clock *B* observes *A* to run relatively slowly. This equal effect is called the reciprocity of observations and results from the fact that each clock moves through the IRF of the other at the same velocity. The shorter period results from the dilation of the time units of the observed clock, such that less of them are counted within a given period than are counted by the observer's clock.

A paradox arises from the fact that each clock is simultaneously observed to be running slow relative to the other clock. Physically, of course, this situation is impossible.

But Special Relativity does not claim to describe a physical reality. In fact SR goes so far as to deny the existance of physical reality. In SR the dimensions of a body are a function of the relative velocity of the observer of the body. Different observers measure different dimensions of the same observed body and these differences may be very large. Special Relativity tells us that no one of these observers is any more special than any other. It follows that where there is no observer a body cannot have dimensions. And if a body has no dimensions then it cannot exist.

The paradox being, of course, that if a body does not exist then it cannot be observed. Thus:-

Special Relativity denies a physical reality.

Non-reality in a Light Ray

Let clock B in the reciprocal clocks experiment emit a light ray towards clock A. According to SR the velocity of the light ray through the IRF of *B*, and also through the IRF of *A*, is 300,000 km/s. But if the relative velocity *v* is acting to separate the two clocks at a rate of, say 200,000 km/s, then how can the physical ray possibly pass both clocks at the same velocity. It is impossible to picture a physical reality of the movement of light through Space which fits with Einstein's postulate.

The only physical reality which can describe this situation is that light maintains a constant velocity through the intervening Aether and the true velocities of the ray through each IRF are the speed of light through the Aether plus or minus (depending on the direction) the Aether velocities of each clock. In this situation light is not isotropic within IRFs.

Under these circumstances the two very different IRF light ray velocities are *measured* by the observers in each IRF to be identical for two reasons. Firstly, the velocity of a one-way passage of a light ray cannot be measured. Only a two way passage of a light ray can be measured which greatly minimises the difference in the measurements of an IRF observer relative to the velocity of the light ray through the Aether, as a fast passage is averaged with a slow passage. Finally the Fitzgerald-Lorenz contraction effect upon the measuring instruments of the observer obscures even that small difference. As a consequence very different light velocities are always *measured* to be equal.

The Twins Paradox

The experiment previously described with two identical clocks reading each other over a separation distance may be modified in such a way as to throw further light on the question of reality.

In this modification clock B passes clock A at near zero distance, at which point the two clocks are synchronised. The constant relative velocity of v is continued for some time until eventually clock B completely reverses its velocity, but still at the same magnitude v, towards clock A. Eventually the two clocks pass each other once more at near zero distance. At this second pass they take readings of each other in order to determine the other clocks measure of the time duration between the two meetings of the clocks compared to their own measure. This experiment is called the Twins or Clock Paradox.

At the end of the experiment it is found that clock B is observed by clock A to be relatively slow. Also, and as would be expected, clock A is observed to be relatively fast by clock B. The necessary time *contraction* of clock A causing clock A to run relatively fast cannot be predicted by Special Relativity. This is a failure of Special Relativity. It is the case that even a single failure of a theory condemns the whole theory.

The fact that two identical clocks show different measures of the time duration between the two meetings while physically stationary and adjacent demonstrates that the time keeping of clocks is a real rather than just an observed phenomenon. Thus physical reality does indeed exist in contravention to the conclusion of Special Relativity.

The Problem of Inertiality
It should be recalled that Special Relativity makes predictions between inertial reference frames.

IRFs, as their name describes, must not be accelerating. Therefore material bodies stationary within an IRF must also be inertial. Thus it is strictly not possible to say that the M-M apparatus existed within an IRF as it is subject to the combined centrifugal accelerations of Earth rotation and Earth's solar orbit. Thus Einstein's solution to the null result of the M-M experiment was not actually applicable to the experiment. The Theory of Special Relativity, should never have got past this initial and most basic point.

One reason why SR has been widely accepted in the scientific community in application to non-inertial bodies is that, at the time and for the next one hundred years, nobody could come up with a better theory, albeit many have tried. G Builder, Herbert Ives and S. J. Prokhovnik are just three scientists who tried valiantly to replace Special Relativity but failed. Time will tell whether the Aether Theory of Velocity Effects will be more successful. But with the advent of a new contender criticisms of the logic of SR once again take on the serious attention which they originally deserved.

The difficulty with the SR requirement for IRFs is that such a perfect state of zero acceleration can never be reached in practice. As a consequence Special Relativity cannot apply to any body in the Universe.

It might be argued that a small degree of acceleration will merely lead to a small difference between the SR prediction and the actual observation, and that this small error may be quite acceptable to the observer. Although a small error might be considered acceptable it must be asked at what point does an acceleration become so large that it becomes unacceptable. There is no theory which may be used to predict velocity effects in highly accelerating bodies other than the Aether Theory of Velocity Effects, which is equally applicable to bodies with zero acceleration.

Again it might be argued that a small degree of error might be acceptable if IRFs were merely employed as mathematical tools within the body of Einstein's theory - but this is not the case. The IRFs are instead employed in the most fundamental part of the theory, the postulates. The SR postulates do not allow for accelerations up to a specific point and no further. They demand absolute zero acceleration at all times.

The Scientific Cost of Special Relativity

For more than a century, and despite the efforts of many different physicists over that period, no Aether based theory of velocity effects could successfully predict relative velocity effects. So although SR was not a perfect theory yet it was substantially better than any of the alternatives. It was only natural under those circumstances that physicists ignored the many difficulties of Einstein's theory in favour of its seemingly successful predictions. The Special Theory of Relativity was therefore considered to be a step forward and, particularly as a consequence of its prediction of the interrelationship between mass and energy, it has been considered not just a step forward but a very great step forward.

Although physics gained a theory for the prediction of relative velocity effects yet a great deal was also lost to physics as a direct consequence of the loss of the Aether. The truth of this statement will become clearer with each chapter of this book.

Chapter 6

The Electric Field

The hypothesis of the Aether requires the electric effect - in what ever form that may take - to be supported by the Aether at each and every point in the Universe. As the magnetic force is generally considered to be a fundamental force it might be thought that the magnetic effect is also directly supported by the Aether.

The total force exerted by a moving charge upon a test charge is currently theorised by modern physicists - as followers of James Clerk Maxwell - to be a complex combination of an electric and a magnetic effect. If, on the otherhand, it is postulated that the magnetic force does not actually exist in a fundamental form, then the electric force must be different in magnitude to that currently accepted in order to give the same total force as is observed. Therefore the determination of the electric field of a charge moving through the Aether depends upon whether it is first assumed that a magnetic field exists or not. The proof of the truth of the postulate then depends upon whether a logical, believable and acceptable physics can be constructed upon the new foundation.

The Background to Magnetism

It is natural to accept that the magnetic force does exist simply because the textbooks and the physics professors tell us that it does. It might also be assumed that magnetism exists simply because magnets exist.

In order to understand current beliefs in electro-magnetism one needs to go back to the beginning when these effects were first observed. The electric force was first noticed when a non-conducting substance, such as amber, was vigorously rubbed with dry fur. Afterwards it was noted that the amber attracted light objects such as small pieces of paper. Hence a new force had been discovered - one which pulled on pieces of paper. The amber was said to be charged with an unknown factor called electricity. It was discovered later that electric charge comes in two forms, the positive and the negative, where like charges repel each other and unlike charges attract each other.

Magnetism was discovered several thousand years ago - possibly by the Chinese - in an oxide of iron called lodestone. Lodestone was found to attract iron and to orientate itself along a line joining the North and South poles of the Earth when free to do so. The end of the lodestone which pointed towards the North Pole was called North and the opposite end was called South. It was discovered that like poles of lodestone magnets repelled each other while opposite poles attracted each other. It was also found that an electric charge, either negative or positive, had no effect whatsoever upon either the North or the South pole of a magnet. Similarly neither magnetic pole had any effect upon a charged object. It appeared that the electric and the magnetic forces were entirely separate in effect and therefore different in origin. So here, it seemed, was a new force of nature - the magnetic force.

The knowledge of magnetism considerably advanced in the 19th century when it was discovered by Hans Christian Oersted in 1820 that the magnetic force could be generated by passing a current of electricity through a wire. The lines of magnetic force - the direction in which tiny test magnets or iron filings align themselves - were found to be circles centred on the current carrying wire. The strength of the magnetic force diminished the greater the distance from the wire. Furthermore if the wire was formed into a round coil then the coil acted like a natural iron magnet, in that one side of the coil became a North pole and the opposite side a South pole. Reversing the current direction reversed the polarities. The similarity between a coil and a magnet is strengthened when the coil is constructed so as to have length. This shape, called a solenoid, acts in a similar manner to a bar magnet. Thus the magnetic force is found to arise from an electric current - in other words from the movement of charged particles (in this case electrons).

Later, ferro-magnetism was also found to be caused by the movement of charged particles. The electrons of iron atoms generate magnetic fields as they move in their orbits around the iron nucleous. The electron also generates a magnetic field as a result of its intrinsic spin. In the non-magnetised material these many small fields all point in different directions and so each small magnet is cancelled out by others pointing in the opposite direction. However the individual fields can be induced, by means of an external magnetic field, to line up in a single direction. Then the individual fields reinforce each other to give the effect of a permanent magnet.

It can be seen that there is a strong connection between the electric and the magnetic forces. The electric force arises from stationary charged particles, whereas the magnetic force arises from the relative movement of charged particles - or more precisely the movement of their electric fields.

The eminent physicist James Clerk Maxwell took this connection between the two forces a stage further. Via his famous Maxwell equations he showed that an electric force field (determined by a relatively stationary observer) would be observed as some combination of an electric and a magnetic field if the observer was moving relative to the field. Similarly a moving observer would view a purely magnetic field (again determined by a relatively stationary observer) as some combination of an electric and a magnetic field. Maxwell also stated that a changing electric field creates a magnetic field and a changing magnetic field creates an electric field. Thus in Modern Physics the two fields are inextricably linked and the two forces appear to have merged into each other to an extent which leaves us unsure as to their true reality. Even the Nobel Prize winner, Richard Feynman, admitted that it was impossible for him to envisage the physical manifestation of the electric or the magnetic force.

This book goes one step further than Maxwell - a big step. It assumes from the very beginning that the magnetic force does not exist. Electric fields are therefore always electric fields whatever their relative velocity to an observer. It follows that the electric force alone must, somehow, explain all of the so-called magnetic effects. How this occurs is described in the chapter on magnetism.

The Structure of the Aether

In a material substance the velocity of acoustic wave propagation is given by $v^2 = C/\rho$ where C is the coefficient of stiffness and ρ the mass density of the material. As the Aether hypothesis states that the velocity of electric waves through the luminiferous Aether is determined by the Aether substance it is not unreasonable to expect that the velocity of propagation is given by an equation of a similar form such that:-

$$c^2 = C_e/\rho_e \tag{6.1}$$

where C_e is the mass-less pressure per unit ·deformation of the Aether and ρ_e is the Aethon density relative to a more fundamental sub-Aether.

The concept of Aethons as identical Aether 'atoms' of which the Aether is constructed is not essential to the theories described in this book but they do assist in an understanding of the processes of the Aether.

So what in our world does Aether mass-less pressure and Aethon density equate to?

It is suggested that Aether pressure is the electric potential and Aether density determines the gravitational potential.

An increase in Aether pressure naturally leads to an increase in Aether density. It is assumed that this relationship is linear such that the ratio of the two factors does not change and hence the propagation velocity does not change.

The Universe is not directly concerned with either the electric or the gravitational potential. But it is very much concerned with the gradient of these potentials. The gradient may be described either locally as the potential difference over an infinitesimal small distance or indeed over a much larger distance where one point is an arbitrarily chosen datum point. An elevated mass-less pressure relative to a datum ambient level equates to one electric polarity while a depressed mass-less pressure equates to the opposite electric potential.

It follows that there must be a maximum possible depression of mass-less pressure as pressure cannot go below zero. Thus there is an upper limit to whichever one of the electric polarities equates to a depressed mass-less pressure. This may possibly give a universal bias favouring particles of one polarity over the other polarity.

The Aether Static Transmission Mechanism (ASTM)
The mechanism of the transmission of the electric potential from a source of elevated potential outward through Space to create the electric potential field is a function of the action of the local mass-less pressure (electric potential) at each Aethon upon its contiguous neighbours. This action is propagated at the speed of light.

It is rather difficult to fully understand how such a mechanism operates between individual Aethons. Instead concentric shells of Aether thickness dr centred on a point source are considered. It is assumed that the magnitude of the electric potential is constant throughout each shell as a consequence of symmetry. In a static electric field (the source is stationary in the Aether) the total 'force' (mass-less pressure times shell surface area) must be equal on either side of each shell. This requirement determines the form of the static field.

Thus, if the shell surface area is termed a then we have:-

$$\phi_2 a_2 = (\phi_1 a_1 + \phi_3 a_3)/2 \qquad (6.2)$$

where shell 2 is sandwiched between shells 1 and 3.
From eqn (6.2) and taking the radius of the shell in question to be r we have:-

$$\phi_2 r^2 = [\phi_1 (r - dr)^2 + \phi_3 (r + dr)^2]/2 \qquad (6.3)$$

Putting ϕ_1 to unity it is found that $\phi_2 = (1 - dr^2/r^2)$ and $\phi_3 = (1 - 4dr^2/r^2)$.

Thus the electric potential diminishes by dr^2/r^2 for each shell relative to the adjacent shell on the source side.

The gradient of the drop over distance dr is therefore dr/r^2.

The integration of this gradient gives $-1/r$.

Thus the magnitude of the electric potential difference diminishes with the inverse of the distance from the source - which is what we expect for a field of electric potential.

Aethon Sources

The source of electric potential difference may be a single Aethon.

It might be thought that only Aethons with an electric potential significantly elevated or depressed with respect to the potential of the surrounding contiguous Aethons should be considered as a source. In fact the magnitude of the difference in potential between a chosen source Aethon and its neighbours is of no significance. Thus every Aethon influences all other Aethons via the ASTM whereby the degree of potential difference diminishes with inverse distance and delayed by the propagation time d/c. Simultaneously every Aethon is influenced by the action of the ASTM mechanism sourced by every other Aethon in the Universe. The result is the ambient electric potential field.

Alternatively, it may be interpreted that each Aethon generates its own separate potential field and the infinity of these fields superposition upon each other to create the ambient field.

The electric fields are, of course, constantly changing as charged particles and electric waves move through the Aether.

Charge Sources

An electric potential source may also be a charged body. In this case the charged body occupies a volume of many Aethons. Nevertheless at a distance from the charge, large relative to its dimensions, that body may be considered to be a theoretical point. A further difference from an Aethon source is that the elevated potential of a charge remains constant - the magnitude of the charge being determined by the potential difference to ambient at unit radius from the source. The ASTM is equally applicable for deriving the potential field surrounding an electric charge as for a single Aethon.

As the electric potentials of a charge are very high the electric fields produced by charges are of considerable significance to the working of the Universe such that the ambient potential field is mostly a function of the effect of charges. But the gradient of the ambient field at any point may not be significantly determined by any particular one of the various charges which contribute to the field.

It can be seen that the electric potential field caused by a charge diminishes with inverse distance from the charge. This result is rather different from the teaching of modern physics which takes the physical electric effect to be the electric force E diminishing with the inverse *square* of the distance from a point source.

The implication arising from the physical electric potential field is that it is the gradient of the electric potential which is the cause of the acceleration of charged particles, as the gradient diminishes with the inverse square of the distance, in keeping with our knowledge of the action of accelerating fields. The direction of acceleration is determined by the direction of the potential gradient. This result is in keeping with the hypothesized internal structure of an FMP as described in the first chapter, in which the asymmetric electric potential geometry of the FMP is modified by the superposition of the gradient of the underlying ambient potential field.

An FMP is not affected by its own gradient so the effective ambient field is that which would have existed in the absence of the FMP in question. For example, where only two charges exist in Space separated by some distance then each charge exists in and therefore is affected by the field of the other charge alone.

It is now apparent that a charge accelerated by the gradient of the ambient electric field experiences the immediate gradient of the field but has not and can have no 'knowledge' whatever of the various sources of that field. Similarly the sources of the ambient field have no knowledge of the effect of their fields on any other body. Thus charged bodies only indirectly create an action of attraction or repulsion upon each other.

Changing Fields

Let us assume that a single charged body is suddenly created at a point in the Aether (charged bodies in reality are only created in oppositely charged pairs). The elevated electric potential of that body will be passed outwards in all directions according to the requirements of the ASTM at the speed of light, so the effect of the potential field can never extend beyond the distance ct where t is the time elapsed since the 'birth' of the charged body. Thus the entire electric field is continuously changing asymptotically towards the inverse distance form at all points, albeit the field may be considered relatively static within a distance from the source charge much less than ct.

The electric field of a charge is not continuously emitted from the source at the speed of light. Instead, it is only the differences with respect to the static field which are propagated. When the source charge is moving through the Aether - as always is the case to one degree or another - then its associated electric field is constantly changing. Although it is only the changes in the field that are propagated outward nevertheless it can still be considered theoretically as though it were the whole field which is being continuously propagated outward.

Once 'emitted' by the charge the electric field propagates through and with respect to the Aether. Consequently a moving charge moves through its own field such that the field is asymmetric with respect to the charge.

Neutral and Complex Particles

Neutral particles are assumed to be constructed of equal positive and negative charges. Thus if positioned on an electric gradient one charge will move one way and the opposite charge will move the other way, but only just enough to re-stabilize the orbiting pair. The net effect is, of course, zero acceleration.

Up to this point the Aether hypothesis has assumed that the acceleration of a charged particle is purely a function of the gradient of the electric field at the location of the particle. But the degree of acceleration is also a function of the mass of the particle. The mass of a body derives from how easily or otherwise the assymetric geometry of a particle is modified by a unit gradient in a unit time. The particle geometry, and hence its mass, may vary from particle to particle. Furthermore particle geometry varies with Aether velocity.

Some charged particles are composites of a number of

fundamental particles. Thus a particle constructed of two negative and one positive FMP has a mass of three units but a net charge of only one negative unit. Although the conglomerate particle would appear to be situated only in the ambient field yet each individual FMP is also situated in the field of its two companions. The oppositely charged FMPs accelerate in opposite directions as a consequence of the ambient field and thus the effect is that the distance between these particles changes to the point where stability is eventually restored in the combined particle fields and the ambient field.

Thus the actual ambient field gradient for each of the three particles is different to the over-all ambient field gradient. The result is that the conglomerate particle accelerates at only one third the rate of a single fundamental particle and so its mass is considered to be three time that of a single particle.

This explanation does not give a precise and quatitive description but merely attempts to provide an understanding of the underlying mechanism of the acceleration of more complex particles.

The Electric Field of a Moving Charge

Consider the moving charge when at point Q on the x axis in Fig. (6.1).

The electric field is effectively emitted at the speed of light by the action of the ASTM in all directions and propagates outward in all directions from point Q. This emission arrives at point $P(x, y, z)$, distance r_o from point Q, after a time delay of r_o/c. In the meantime the charge has moved along the x axis at Aether velocity v by the distance $r_o v/c$ to point O - which is taken to be the origin of the x axis. There are, of course, an infinity of points P which lie on the

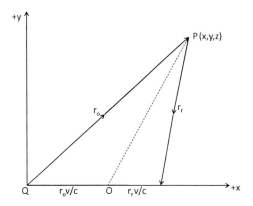

Figure 6.1: The Electric Field

surface of a sphere of radius r_o centred on point Q.

From Fig.(6.1) and by Pythagorus we have

$$r_o^2 = \left[(r_o v/c + x)^2 + y^2 + z^2 \right]$$

Hence:-

$$r_o = \gamma \left[\gamma x v/c + \sqrt{\gamma^2 x^2 + y^2 + z^2} \right] \qquad (6.4)$$

where $\gamma = 1/\sqrt{1 - v^2/c^2}$, the Lorenz function.

The equation for the electric field of a stationary charge is:-

$$\phi = q/4\pi e_0 . r_o$$

Thus the field of a moving charge is given by:-

$$\phi = q/4\pi e_0 \gamma \left[\gamma x v/c + \sqrt{\gamma^2 x^2 + y^2 + z^2} \right] \qquad (6.5)$$

It can be seen that the potential field is not symmetrical along the x axis relative to the instantaneous position of the charge at O. Instead it is shifted backwards along the x axis in the x direction by the factor $\gamma x v/c$.

It can also be seen that the x axis is effectively contracted relative to the y and the z axes by the factor $1/\gamma$ such that $x' = \gamma x$, where x' is the axis in the IRF of the moving charge parallel to the x axis in the Aether. The factor γ in front of the expression in square brackets modifies all three IRF axes equally. An equal modification to distance in all directions is considered to be a modification to the time dimension in the IRF of the moving charge. This is because time is taken to be an artificial dimension given by d/c where d is distance in any direction in the IRF in question.

This description of the electric field is quite different from that accepted by the followers of Maxwell (modern physics). One cause of this difference is that modern physics considers that charge is a substance and that substance occupies volume. The Lienard-Weichardt potentials stem from the assumption that charge is a volume occupying substance and the application of these potentials, together with the assumption that magnetism is a fundamental force of nature, leads to Maxwell's different conclusion for the electric field equation.

Communication back from the Field

When the electric potential arising from the charge reaches point P the actual charge is at point O. If the charge Q while at point O wishes to observe the effect of its own field at point P in accordance with the Two-Way Maxim - for example upon a test charge at point P - it has no means of directly doing so. Any communication from point P to point O will take time and in the meantime the observer has moved on from point O. Thus the charge can never truly observe its own field. The act of observation intrinsically modifies the appearance of the field. One example of 'observation' is by means of the field of an affected test charge at point P which in turn operates at the speed of light upon the charge Q via its own electric field.

In Fig.(6.1), a return communication at the speed of light is shown from point P which reaches the charge Q at point 2. The equation for the return distance r_r employing the same principles as used for the outward journey, is given as:-

$$r_r = \gamma\left[-\gamma xv/c + \sqrt{\gamma^2 x^2 + y^2 + z^2}\right] \qquad (6.6)$$

From the standpoint of the moving charge a signal was emitted from the charge which reflected from an object at point P and returned to the charge. During this period and by definition the charge has not moved in its own IRF. The distance $+\gamma xv/c$ in eqn. (6.4) is therefore cancelled out by the distance $-\gamma xv/c$ in eqn. (6.6) over the full out and return communication.

Therefore the equation for an *observed* electric field moving with respect to the observer is:-

$$\phi = q/4\pi e_0 \gamma \sqrt{\gamma^2 x^2 + y^2 + z^2} \qquad (6.7)$$

It can be seen that the observed electric field is symmetric, contracted in the x direction with respect to the other two axes and time dilated.

An observer stationary with respect to the field does not actually notice this contraction as he possesses no measure of distance which is more fundamental against which he might make a comparison. Thus an observer stationary with respect to the charge assumes the electric field to be perfectly spherical around a charge irrespective of his absolute velocity.

This description of the observed electric field is different to the Maxwell derived equation of the field, given as:-

$$\phi = q\gamma/4\pi e_0 \sqrt{\gamma^2(x - vt)^2 + y^2 + z^2}$$

The Lorenz Transforms

It has been mentioned that the Lorenz Transforms are equations employed in Special Relativity to transform the dimensions of length, time and mass from one IRF to the other. The Lorenz Transforms are, in fact, equally applicable to Aether theory and are more directly and arguably less questionably derived from eqns. (6.4) and (6.6) above.

The Lorenz Transforms involve the parameter of time whereas eqns. (6.4) and (6.6) do not, but as previously defined $t = d/c$ where d is distance in any direction.

The Aether dimensions are given as x, y, z and t and the equivalent IRF dimensions are ξ, η, ζ and τ. Axes x and ξ are in line with each other.

In eqn. (6.4) the content of the square bracket may be modified by setting $y = z = 0$ to obtain the equation for ξ. ξ is then given relative to the y and z dimensions. Thus:-

$$\xi = \gamma(x - vt) \qquad\qquad (6.8)$$

where $t = x/c$.
Eqn. (6.8) is the Lorenz Transform for the x axis.
By definition we have:-

$$\eta = y \qquad \zeta = z$$

Dividing eqn.(6.4) throughout by c after putting $y = z = 0$ gives

$$t = \gamma(\tau + x'v/c^2)$$

where $x' = \gamma x$ and τ is x'/c.

By substituting eqn. (6.8) we obtain

$$\tau = \gamma(t - xv/c^2) \qquad (6.9)$$

which is the Lorenz Transform eqn for time.

The Lorenz Transform equation for time shows that when a speed of light communication is made - such as in the synchronisation of a distant clock - there is a time 'error' factor of xv/c^2, which we may call the Time Separation Factor (TSF), in the communication. The TSF is due to the fact that the observer assumes from measurement evidence that the one way speed of light through his IRF is c when, due to his Aether velocity, it is not. The TSF does not however affect the reading or measurement of any body as the necessary return communication effectively puts the TSF to zero by effectively putting distance x back to zero. Thus the TSF is generally of no consequence in communication between bodies.

It should be noted that the Lorenz Transforms, as calculated so far, operate only on the dimensions of the electric field from which they are derived, between a field stationary in the Aether and one moving through the Aether. It is later demonstrated that the dimensions of matter are affected to an equal degree as a direct result of these changed dimensions of the electric field.

An essential requirement for the development of Aether theory is the generation of the Lorenz Transform equations for velocities from the Lorenz Transforms for distance and time. This is a standard mathematical exercise repeated in Appendix 4. The resultant equation is shown here:-

$$\gamma_u/\gamma_u' = \gamma_v(1 + u_x'v/c^2) \qquad (6.10)$$

This Velocity transform equation is applicable to any factor or phenomenon which is a function of the factor γ, such as the electric field.

Chapter 7

The Aether Theory of Velocity Effects

The Electric Field and Matter

Consider an atom moving through the Aether. The positive electric field emanating from the nucleus will be contracted in the direction of atomic movement in accordance with the new Aether electric field equations derived in the previous chapter. As the atomic electrons orbit the nucleus under the accelerating attraction of the nuclear electric field it follows that the orbits of the atomic electrons must be modified to a certain degree.

J. S. Bell - well known in the field of quantum mechanics - considered this matter in his paper entitled 'How to teach Special Relativity'. But don't let the title concern you. Bell calculated the change in the orbit of the single electron in a hydrogen atom in response to the velocity modified electromagnetic fields of the nucleus, where the velocity was relative to any chosen IRF. Thus Bell's calulations are equally valid for the reference frame of the Aether. Bell considered the forces on the electron to be a composition of the electric

and the magnetic, but the total force must be independent
of theories as to its origin. Hence Bell's calculations are
equally applicable to Aether theory where magnetism does
not exist. Bell's conclusion was that the electron orbits were
compressed in the direction of velocity by the factor $1/\gamma_e$ -
which is to exactly the same extent as the compression of
the electric field.

As the outer electron orbit of an atom effectively determines the dimensions of that atom and as bulk matter
is constructed of atoms then the dimensions of all matter
are determined by the outer electron orbit. Thus all matter
is contracted in the direction of Aether motion by the factor $1/\gamma_e$. In other words the Fitzgerald-Lorenz contraction
effect is indeed a real physical effect as suggested by its pro-
posers which is soundly based upon the Aether hypothesis.

Thus the contours of electric field strength surrounding
a charged body and measured by means of standard matter
rods employed by an observer co-moving with that body will
determine that the contours are perfectly spherical in shape
irrespective of the Aether velocity of the charged body and
the true deformation of the field. This is simply because the
electric field and the measuring rods are equally contracted
at all Aether velocities andin all directions.

Bell also found from his calculations that the period of
the orbit of the atomic electrons dilated according to the
factor γ_e. Once again this is to exactly the same extent as
time dilation occurs in the electric field. The orbit of the
electron is just one example of a material time unit generator, but it is only a trivial extension to accept that all time
unit generators - and hence time in general for matter sys-
tems - are equally affected by Aether velocity induced time
dilation as the particular example of the electron orbit.

Hence Bell's work demonstrates that Aether velocity effects of length contraction and time dilation on matter are real physical effects independent of observation.

The Lorenz Transforms and the consequent various velocity effects are therefore equally applicable to matter moving through the Aether as they are to the electric field.

This is the Theory of Absolute Velocity Effects

The Meaning of an IRF

As every material body suffers velocity effects as a function of its Aether velocity it follows that the measuring rods and clocks of a body - for example, here on Earth - suffer a degree of length contraction and time dilation. An observer on Earth is thus forced to measure the Universe with distorted instrumentation. As a consequence he will necessarily obtain a distorted picture of the Universe. That distorted picture is the observer's inertial reference frame.

As each body has its own particular absolute velocity then every body sees and measures the Universe differently to any other body. That is the reason why IRFs are particular to the observer. For example, the measured distances to particular distant stars and galaxies will be different for observers moving at different Aether velocities even though the observers occupy (temporarily) the same point in Space.

The IRF Transformation of Mass

The Lorenz Velocity Transforms of length contraction and time dilation now applicable to matter generally also provide the consequence that mass increases as a function of velocity. Thus:-

$$m = \gamma m_0 \qquad (7.1)$$

where m_0 is the mass of a body stationary in the Aether.

The derivation of the mass increase equation arises from

a consideration of the conservation of momentum together
with the employment of the Lorenz Transform equations for
distance, time and velocity. As this derivation is a standard
work the details of the calculation are not repeated here.

The Aether Theory of Relative Velocity Effects

The velocity effects which can be observed and measured are
those of the dimensions of bodies moving at known relative
velocities to the observer. All experiments and observations
to date show that a body A observing a second body B moving at relative velocity will observe length contraction, time
dilation and mass increase in that body - all as a function
of relative velocity.

For the sake of staying purely with the physics of velocity effects we temporarily ignore the practical difficulties
of measuring these parameters over distance and at velocity
and merely assume that the feat can be readily achieved by
one means or other.

When an observer measures, say, the length of an iron
bar, he does so by effectively counting how many of his standard length units make up the length to be measured. If
the iron bar is ten meters long and the standard length is
one meter then the observer will count ten standard units.
The measurement operation is to divide the length to be
measured by the standard measure. If, for example, the observer's standard measure was shrunk through velocity effect
to one half meter and the iron bar remained constant then
the measure would give an answer of twenty. Therefore, in
order to check for any changes between the standard unit
measures employed by two different bodies we need to observe the size of the standard unit of the observed body and
divide it by the size of the observer's standard unit.

As we are now aware of the Aether velocity functions

which affect the size of each measurement unit we only need
to divide the relevant velocity functions particular to each
body in order to determine the velocity distortion ratio. In
the case of velocity effects on time and mass the function is
γ_e and for length $1/\gamma_e$. Thus a measure of time and mass
of observed body B by observer body A is modified by the
ratio γ_{eB}/γ_{eA} (inverted in the case of length). This ratio is
called the Velocity Ratio.

Because γ_{eB}/γ_{eA} involves unknown and unknowable Aether
velocities it might appear that the Velocity Ratio can never
be calculated. However the Velocity Ratio can be simply
converted by employing the Lorenz Velocity Transform eqn.(13.11).
This equation requires that the observer A is inertial but the
observed body B may be accelerating.

In the Velocity Transform eqn. $u = V_{BE}$, $u' = V_{BI}$ and
$v = V_{IE}$ where I is the IRF of body A. The Velocity Ratio is
now given by γ_{BE}/γ_{IE}. Thus, with a small re-arrangement,
we have:-

$$\gamma_{BE}/\gamma_{IE} = \gamma_{BI}(1 + V_{BIx}V_{IE}/c^2) \qquad (7.2)$$

V_{IE} is the Aether velocity of both observer A and frame I.
The factor γ_{BI} is the velocity effect function involving the
relative velocity V_{BI} between the two bodies.

Eqn.(7.2) states that the Real Ratio is given by the relative velocity effect function multiplied by the factor $(1 +
V_{BIx}V_{IE}/c^2)$. Over a measurement time period T this fac-
tor goes to $T + xV_{IE}/c^2$ where distance x is the distance
moved over the measurement period by the observed body
in line with the Aether velocity of the observer. xV_{IE}/c^2 is
the Time Separation Factor. The return of an observation

to the observer at the end of the measurement period, by any means whatever, satisfies the Two-way Axiom and effectively puts distance x, and hence the TSF, to zero giving the result that the observed Velocity Ratio equals:-

$$\gamma_v T$$

Alternatively, substituting the modified Lorenz Transform for time,
$t = \tau \gamma_{IE}(1 + V_{BIx}V_{IE}/c^2)$ into eqn.7.2 and taking body B to be any body generally gives:-

$$\gamma_{BE}/t = \gamma_{BI}/\tau \tag{7.3}$$

which states that the ratio of the velocity effects on a given body to the units of any IRF are a constant.

Thus we may derive an alternative expression for the Real Ratio:-

$$\gamma_{BE}/\gamma_{IE} = \frac{\gamma_{BI}}{\tau}\frac{t}{\gamma_{IE}} = \gamma_R/k \tag{7.4}$$

which states that the Real Ratio is given by the relative velocity effect on a body moving through an IRF, measured in the units of that IRF, and divided by a constant factor γ_{IE}/t for that particular IRF. This constant may be taken to equal unity as the observed velocity effects on a relatively stationary body are zero.

Both eqns. (7.2) and (7.4) describes the Aether Theory of Relative Velocity Effects (ATVE), but from slightly different viewpoints.

It is not immediately obvious that the measurement of length and mass over a separation distance requires a time period to carry out that operation. The distance from the observer to a point on a body may be determined from one half the total duration of a two way light pulse reflected from that point back to the observer. The length of a body may be determined from the difference between two such measurements to points at each end of the length to be measured. The subtended angle between the two points must also be measured so as to determine the attitude of the measured length to the observer.

Mass can be measured at distance by observing its acceleration in a force field of known strength. Over a given time period t either the change in velocity ($m = Ft/v$) or the change in distance moved ($m = Ft^2/2d$) may be measured in order to determine the mass.

Over a measurement time period of T the factor on the RH side of eqn.(7.2) goes to ($T + xV_{IE}/c^2$). So, irrespective of the movement velocity the time factor xV_{IE}/c^2 (the TSF) acts as a consequence purely of the physical separation of two points in an IRF irrespective of whether the distance is traversed by a matter body, by an electric field or by a light ray. This explains why the movement of the timing clock from one end of the race-track to the other in the attempt to measure the one way speed of light must fail, for the clock movement has exactly the same time effect as the passage of a light ray between the same spatial points.

Pseudo- Aethers or IRFs

The measurement of matter bodies in terms of their length, time and mass - and indeed all other parameters - are found to be functions of velocity through the IRF of the observer when measured in the time units of that IRF. This means that each IRF is effectively acting as a pseudo-Aether. This is the case to the extent that the speed of light through an IRF may be considered to be constant and isotropic, even though in actuality it is not. The principle of the *observed* invariance of the laws of nature between IRFs therefore holds true. But the underlying reasons for the observed invariance - the real changes in time and distance units - can now be appreciated.

Relative Velocity Observations Viewed from the Aether Frame

It is instructive to consider measurements within an IRF from the point of view of what is really happening in absolute Aether terms.

It can be seen that the observation of a velocity effect using the ATVE is the combined result of three separate effects; the real Aether velocity effect upon the observed body, the real Aether velocity effect upon the observer and the Separation Factor, $(1 + V_{BIx}V_{IE}/c^2)$. Thus:-

$$\gamma_{BI} = \gamma_{BE}/(1 + V_{BIx}V_{IE}/c^2)\gamma_{IE} \qquad (7.5)$$

The significance of the Separation Factor relative to the other two effects depends upon the Aether velocity of the observer relative to the Aether velocity of the observed body. This can be more easily appreciated if the situation is simplified somewhat. We can assume that the Aether velocities of both bodies are small relative to c and that both velocities are in the line joining the two bodies. The actual separation distance is immaterial. Thus the relative velocity V_{BI} (which now equals V_{BIx}) between the bodies is now $V_B - V_A$ as seen by A and $V_A - V_B$ as seen by B.

We will assume that the two bodies are identical clocks and we are measuring the ratio of their time units. By binomial expansion on eqn. (7.5) the total fractional difference observed by clock A is:-

$$(T_A - T_B)/T_A = [V_B^2 - V_A^2 - 2(V_B - V_A)V_A]/2c^2 \qquad (7.6)$$

and for clock B it is:-

$$(T_B - T_A)/T_B = [V_A^2 - V_B^2 - 2(V_A - V_B)V_B]/2c^2 \quad (7.7)$$

It is easy to show that both the above equations simplify as expected to $V_r^2/2c^2$ - which is the prediction of both the SR and the ATVE final equations.

Now in eqn.(7.6) consider the ratio of the real effect to the Separation Factor.

$$(V_B^2 - V_A^2)/2V_A(V_B - V_A) = (V_B + V_A)/2V_A \quad (7.8)$$

If V_B equals V_A then this ratio is unity. If V_B is greater than V_A the ratio of real effect to Separation Factor is greater than unity and can approach infinity with high values of V_B. On the other hand if V_B is less than V_A the ratio is less than unity and the Separation Factor exceeds the real effects. The Separation Factor can be as large as twice the magnitude of the largest Aether velocity effect upon either clock A or B.

The Separation Factor is therefore of great significance particularly in the case of the Twins Paradox type of experiment where the observers IRF changes to a different IRF at the reversal of the travelling clock.

If the observer is not stationary in the Aether then the total observed velocity effect is always a mix of real effect and Separation Factor - the exact mix being dependant on the relative Aether velocities of the two clocks.

The Advantages of the ATVE over SR

One obvious advantage of the ATVE is that matter bodies are real with dimensions independent of observation. Furthermore the observations of relative velocity effects can be predicted between bodies neither of which need be inertial.

A change of the observer's IRF between observations can be allowed for, the explanation being given in the next chapter. A consequence of this ability is that the Twins and the Dingle Paradoxes are no longer paradoxes.

The major advantage of the ATVE is that the Aether and its consequent explanations are brought back to Physics. Material bodies are real and Space is a substance which determines the separation of the heavenly bodies and, via its natural propagation velocity, determines the local speed of light. The Universe is not only more readily pictured and hence more easily understandable but in real terms it is actually simpler.

Chapter 8

The Twins Paradox

The Twins Paradox Class of Experiment
This book makes no apologies for devoting a whole chapter to the Twins Paradox type of experiment for the simple reason that it is this experiment which most clearly uncovers the failings of Special Relativity. This experiment highlights the major differences between the Aether Theory of Velocity Effects and Special Relativity.

The Twins Paradox group of experiments are experiments upon the phenomenon of time dilation as a function of absolute and relative velocity and are therefore experiments made upon clocks in various situations. The big advantage of experimenting upon time dilation rather than length contraction or the increase in mass effect is that clocks record the effect of time dilation as a slowing of the time-keeping of the clock, which even if the slowing was for only a short period, that clock remains slow even when it returns to 'correct' time-keeping. Clocks therefore act as a form of storage device for periods of time dilation. By this means the effects of time dilation can be observed at a later time when observational effects over distance, which otherwise might modify the clock reading, are absent.

The essence of the Twins or the Clock Paradox, as it should really be known, is the journey of one clock B (called the travelling clock) at a relative velocity with respect to an identical and stationary (inertial) clock A, large enough to cause a measurable time dilation effect. The journey of clock B starts from a point in the IRF of clock A and subsequently finishes at the same point. The exact route and velocity of the journey is not of importance. However certain types of route and constant velocities are normally chosen purely to simplify the mathematical calculations of the time dilation effect. For example, the journey may be outwards in a straight line at a certain constant velocity while the return journey is also in a straight line at the same velocity. For the same reason of simplicity the clock journey might be chosen to be a perfect circle at a constant orbital velocity.

The Twins Paradox experiment was first proposed by Langevin in 1911, just 6 years after the publishing of Special Relativity, as a test of that theory. He described his experiment using identical human twins as a form of identical clocks. One of the twins took off from Earth in a rocket ship leaving his other twin stationary on earth. At a certain velocity the rocket motors were switched off and the rocket continued away from Earth at a constant velocity (ignoring the pull of Earth's gravity). After a certain time, in this case measured in years, the rocket ship quickly reversed direction and travelled back to Earth, once again at the same constant velocity. A final deceleration placed the traveller upon Earth by the side of his stationary twin. The twins then compared their apparent ages in order to determine whether the travelling twin had experienced time dilation and, as a consequence, aged more slowly than his stationary twin.

But from now on, both for convenience and for accuracy, only theoretically perfect clocks will be employed in this discussion of the Twins/Clock Paradox. It is also assumed that each clock may act as an observer of the other with the means to read the other clock over any separation distance using the two-way interrogation method.

Special Relativity predicts that clock A (the inertial clock) will observe that clock B runs relatively slowly equally on both the outward and the return journeys. The direction of velocity through the observer's IRF has no significance as the factor involved entails the square of the velocity. Consequently, at the final reunion of the two clocks, A will read B as slow. It therefore follows that we should expect that B would read A as being fast. In real experiments of this type observations do indeed agree with these predictions.

But there is an alternative way of looking at the experiment. A prediction of time dilation may equally be made by clock B of clock A, for B's observations during both the out and return journeys must equally be governed by the predictions of SR. From the point of view of clock B he sees A moving in his IRF. This is equally valid in the context of SR as the IRF of B has equal significance to that of A, although in B's case he occupies two different IRFs - one for the outward journey and one for the return journey. Thus B's SR predictions of the final reading of clock A are equally as valid as A's predictions of B. Of course there could be a special effect ocurring at the change of IRFs by clock B. However, SR offers no prediction for that particular circumstance.

According to SR, clock B predicts that clock A runs slow on both outward and return journeys. As a consequence B expects to read clock A as slow at their final re-union. Yet actual observation indicates the opposite - that B will read clock A as being fast. Hence the SR predictions of clock A

upon B agree with observation but the SR predictions of
clock B upon A do not agree with observation.

This result is euphemistically called a paradox - the Clock
or Twins Paradox. However it is simply an example of the
failure of the Special Theory of Relativity.

SR Explanations of the Clock Paradox

The most frequent response to the Clock Paradox by the
supporters of Special Relativity is that as there are periods
of acceleration of the travelling clock during the experiment
and these accelerations, particularly the reversal of the travelling clock, being outside of the compass of SR lead somehow to the false prediction of clock B. This statement is
made notwithstanding that the supporters of SR are perfectly content to make and to trust the application of SR to
the accelerating platform of the Earth for all other observations of velocity effects.

In the most common form of the experiment the travelling clock starts from a stationary position next to clock
A, at which point synchronization of the two clocks takes
place. Then clock B accelerates away up to a constant velocity followed by a further acceleration at the reversal of
the travelling clock. A further deceleration takes place prior
to the re-union at the end of the experiment when finally
both clocks read each other once again over zero distance
and at zero velocity. Thus there are actually three periods
of acceleration in the experiment as described by Langevin,
one at the start, one at the end and one in the middle at the
velocity reversal.

Two periods of acceleration can be quite easily removed
from the experiment by the following method. Prior to the
start, clock B is already at the experiment's constant relative velocity V and approaching clock A, passes at near

zero distance. Synchronization occurs at near spatial coincidence, and then B continues on its journey away from A.
After the velocity reversal clock B approaches A from the
opposite direction. The two clocks then eventually read each
other at the point of second spatial coincidence. At spatial
coincidence, despite the relative velocity, no observational
effects can occur.

The period of acceleration at the reversal of clock B is
slightly more complex to remove. However, this may be
achieved by the use of a third identical clock, clock C. At
the 'reversal' point clock B is passed at near zero distance by
clock C approaching clock A at the same constant velocity as
B is leaving A. At the passing point clock C is synchronized
to clock B. Thus clock C returns to clock A just as though
it were clock B. At all times throughout this version of the
experiment all three clocks remain inertial. This version of
the experiment is called the Three Clock Paradox.

However, even if the reversal acceleration is not removed
by the employment of the Three Clock Paradox version it
is never possible for the effects of the reversal acceleration
to reverse the final observation of clock B upon clock A, as
is required for SR predictions to agree with actual observation. To achieve this the reversal acceleration would need to
generate a slowing effect of *exactly* twice the observed velocity effect. But there is no connection whatever between the
period and the intensity of the reversal acceleration and the
duration of the constant velocity parts of the journey, so the
two effects are most unlikely to be equal and opposite.

Many people in their attempt to make the Special Relativity prediction agree with actual observation for both clock
B as well as for clock A resort to employing the Lorenz
Transform equations directly, rather than the resultant equation for time dilation. They argue that it must be the ap-

plication of the simple time dilation equation to the Clock Paradox which is at fault. But by employing the Lorenz Transforms they introduce considerable scope for incorrectly arriving at the prediction which they would like to see.

As the SR predictions of clock A entirely agree with observation there is absolutely no need to consider the observations of clock A. Nevertheless the followers of Einstein frequently attempt to consider the observations of clock B by somehow simultaneously involving the observations of clock A. Commonly they bring the separation distance into their calculations and frequently they determine this distance as viewed by A but employed by B. It is not science but it has the advantage to them of giving them the answer they are looking for. For no-one likes to look a fool for having believed in a faulty theory.

The over-riding point of importance here is that when the two clocks are both in the constant velocity parts of the experiment then the relative velocity must be identical for both clocks. Thus each clock, according to SR, must observe an identical effect of time dilation in the other clock. And time dilation may be observed without any knowledge of the intervening distance (although an allowance must be made by the use of a two-way observation for the changing separation distance over the measurement period).

The Circling Clock version of the Twins Paradox

This version of the Clock Paradox is interesting as it throws more light on the essence of velocity effects.

In the circling clock version of the Clock Paradox clock A remains inertial while clock B orbits clock A at a constant radius and at a constant orbital velocity. The start point, which is also the finish point, may be anywhere on the orbit in the IRF of clock A. Consequently the 'Clock Paradox'

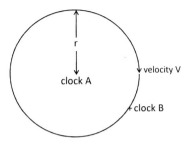

Figure 8.1: The Circling Clock Paradox

journey is one whole orbit.

In the circling clock version clock B is continuously accelerating. Consequently it might reasonably be said that SR cannot apply either to the observations of clock A or clock B. But for the present this point will be put to one side.

Now observations of clocks circling the Earth in satellites show that orbiting clocks do run relatively slowly as a consequence of their orbital velocity. Thus clock A will actually read clock B to be running relatively slowly, and therefore clock B should observe clock A to be running relatively fast. This result is independent of the position of clock A with respect to the orbit of clock B, for clock A may be positioned anywhere within the orbit of clock B, or indeed outside of it. Clock A is merely placed at the orbit centre for the sake of simplicity. The observations between the two clocks apply to the observation of a complete orbit as previously stated but they also equally apply to any part section of the orbit once communication time effects over the differing distances to the start and finish positions are allowed for.

Basically, the result of the Circling Clock Paradox is identical to the standard out and return version of the Clock

Paradox in that clock B cannot make an SR prediction which agrees with the observation that clock A runs relatively fast.

However there is an addition to the Circling Clock experiment which is rather intriguing. In this theoretical version clock B orbits the inertial clock A at constant velocity V on the end of an infinitely light tether of finite length. The tether is then lengthened, but the orbital velocity of B is kept at V. This change in no way modifies the relative timekeeping of the two clocks from velocity effects alone. The only effective change is to reduce the acceleration of clock B. The two clocks continue to take readings of each other over arbitrary time intervals and over the separation distance.

The tether is next lengthened to a great distance, but still keeping the orbital velocity of B constant at V, such that the acceleration on clock B caused by the pull of the tether is now very close to zero. The relative timekeeping of the two clocks will still continue to remain the same.

Now for the crucial part - the tether is cut. Immediately the minute acceleration of B goes to precisely zero. Now both clocks are inertial and SR is instantly in a position to make predictions by both clocks of each other, whereas before the cut it was theoretically unable to do so. According to SR clock B will now predict that clock A is running relatively slowly. Yet prior to the cut clock A was actually running relatively fast. Yet all that has changed is an insignificant degree of acceleration of clock B.

The correct theory of velocity effects must be capable of explaining why the observations of clock B instantly reverse as the consequence of such a minute change of acceleration.

The ATVE Explanation of the Clock Paradox
In the Aether theory real changes in the time keeping of clocks are a function of their absolute velocity. It is the

case that in the standard Clock Paradox experiment clock B must, on one of its journeys, have an Aether velocity which is greater than that of clock A. On the other journey the Aether velocity of clock B will be smaller than that of clock A. So on one journey clock B will run slow relative to clock A and on the other it will run fast relative to clock A. It might be suggested that clock B is observed by clock A so as to see on which journey that clock runs fast and when it runs slow. However, both SR and the ATVE state that no difference can be observed. This because of the effect of the Time Separation Factor acting through the need to observe over a separation distance.

An examination of clock $B's$ journey can be made from the viewpoint of the Aether. The combined effect of the out and return journey can be obtained in a consideration of the time dilation equation, $t = t_0/\sqrt{1 - V_e^2/c^2}$.

In the interests of simplicity we may reasonably consider that the two clocks only move at relatively small fractions of the speed of light so that the difference in time keeping relative to a clock stationary in the Aether over a journey time of T is given by $TV_e^2/2c^2$. We may also take the unknown Aether velocity of clock A to be V_A while the known velocity of clock B through the IRF of A is V. Velocity V may be in any direction relative to V_A but it can be broken down into components in line with and orthogonal to V_A, which we may call V_x and V_y respectively. Thus $V^2 = V_x^2 + V_y^2$. The Aether velocity squared of clock B now equals:-

$$V_B^2 = (V_A + V_x)^2 + V_y^2$$

where V_x may be either positive or negative depending on its direction relative to V_A.
Subtracting the factor V_A^2, as we are only interested in time

differences relative to clock A, gives a time difference for the outward journey of:-

$$T[V_B^2 - V_A^2]/2c^2 = T[V_A V_x/c^2 + V^2/2c^2] = xV_A/c^2 + dV/2c^2$$
$$(8.1)$$

for a total journey path of length d. The factor xV_A/c^2 is the Time Separation Factor (TSF).

The return of clock B to clock A gives $-xV_A/c^2 + dV/2c^2$. Thus the observed time difference at the reunion of the clocks is dV/c^2 as expected. If instead clock B is read by light ray over distance d then the time effect of the observation is just $-xV_A/c^2$, to give a total reading difference of just $dV/2c^2$.

As the velocity V may be in any direction to the Aether velocity of A it may therefore change direction throughout the experiment. Thus the path of clock B may take any shape or form. The total degree of time dilation is given by the integral of the time dilation over each small section δd of the path. Thus both velocity and direction may be continuously changing.

The Clock Paradox is an example of an out and return movement but where the movement is made by matter transport in both directions. However the need to observe the clock reading by any means whatever ensures that the TSF goes to zero. Hence the observed time dilation is a function of relative velocity despite the fact that it originates from absolute velocity effects.

The Effect of Changing IRFs

Just prior to the reversal of clock B in the Clock Paradox experiment the Time Separation Factor is given by xV_{B1}/c^2. Just after the reversal it is given by xV_{B2}/c^2. If the reversal is undertaken in a time which is small compared to the outward journey time of clock B then the separation distance of the two clocks in the x direction may be considered to be constant. Thus the immediate difference in the observation effect of clock B observing clock A over the reversal is $x(V_{B1} - V_{B2})/c^2$.

The calculations required to obtain $V_{B1} - V_{B2}$ are facilitated by a transformation equation between the Aether velocities V_A and V_B. These equations derive from the vector addition of the relative velocity V to the Aether velocity of body $A, (V_A)$ and also to the Aether velocity of body $B, (V_B)$.

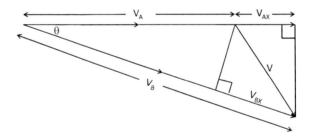

Figure 8.2: Two Clock Vector Diagram

Velocity V may be in any direction to the Aether velocities V_A and V_B. The velocities are taken as fractions of the speed of light. Thus:-

$$V_B^2 = V_A^2 + 2V_A V_{Ax} + V^2$$

and

$$V_A^2 = V_B^2 + 2V_B V_{Bx} + V^2$$

V_{Ax} is the relative velocity component in line with V_A and

V_{Bx} is that in line with V_B. Thus the two x directions are different. By addition we obtain the transformation equation:-

$$V^2 + V_A V_{Ax} - V_B V_{Bx} = 0 \qquad (8.2)$$

Distance x in the TSF can be converted to TV_x, thus for clock B at the reversal point:-

$$TSF_B = TV_B V_{Bx}/c^2$$

Translated by eqn. (8.2) into the constant Aether velocity V_A gives

$$TSF_B = T(V^2 + V_A V_{Ax})/c^2$$

As $T = x/V_x = d/V$ we have

$$TSF_B = dV/c^2 + xV_A/c^2$$

The difference between the two TSFs, immediately before and after the reversal, where xV_A is constant, is therefore:-

$$TSF_{diff} = d[V - (-V)]/c^2 = 2dV/c^2 \qquad (8.3)$$

Thus the observed time of clock A as read by clock B will change by the quantum amount $2dV/c^2$ when observed immediately either side of the velocity reversal of B. Clock A will appear to observer B to have moved a quantum jump faster by exactly the amount required to turn the predicted slow reading at reunion into the actual fast reading of clock A.

So there is a reversal effect, but it is not a function of acceleration as is sometimes mooted but a function of the change in the Aether (and hence also relative) velocity of observer B *coupled with the separation distance from clock A.*

It is worth looking at the continuous reading of clock A by clock B throughout the whole experiment. Over the whole outward journey and up to the start of the reversal (but not including it) clock B sees clock A run slow by the amount $-dV/2c^2$. Immediately after the reversal, with the addition of the reversal effect, clock B now sees clock A fast by the amount $3dV/2c^2$. Then over the return journey clock A appears to run slow by a further amount of $-dV/2c^2$, giving a final total of fast by dV/c^2, which is in exact agreement with observation.

Thus the ATVE makes the correct predictions of the Clock Paradox over the course of the experiment *by each clock of the other at any point in the journey.*

Chapter 9

Magnetism

In the chapter on the electric field it was assumed that magnetism does not exist as a fundamental force of nature. It is therefore necessary to explain 'magnetic' phenomena by other means.

'Magnetic' effects are found to result from the relative movement of electric fields, the commonest example of which is given by the movement of current electrons in a conducting wire. With an electric current the negative electric charge of the electrons is expected to be cancelled by the positive charge of the wire ions. Any forces experienced outside of the wire are considered to be magnetic in nature.

Consider a long straight wire through which an electric current is flowing. It is found by observation that a test charge moving in any direction through the plane of the wire suffers a force at right angles to its direction of motion. The strength of this force is a function of the current in the wire, the relative velocity between wire and test charge, the inverse distance between wire and test charge and the magnitude of the test charge. This force is intuitively unexpected as it might reasonably be thought that the electric force fields of the current electrons and the ions of the wire would exactly cancel each other under all circumstances. It is also unexpected that the net force encountered should be a function of the relative velocity between the wire and the observer. It is therefore not at all surprising that this phenomenon should have been considered a separate force to the electric force.

In considering the operation of the 'magnetic' force we may take the test charge to be the observer (of the force exerted upon itself) stationary in his own IRF. The conducting wire now moves through the IRF of the test charge. This is a reversal of the intuitive way of looking at the apparatus, although whether the test charge or the wire is taken to be stationary is of no significance as it is the relative velocity which matters.

Now the velocity of the current electrons of the wire through the IRF is a combination of two velocities. Firstly, with no current flowing, the current electrons possess the relative velocity of the wire. Additionally they possess a velocity along the wire when a current is flowing. The velocities of the current electrons and the wire ions modify the field of those charges as described by the Aether equation for the electric field, eqn.(6.7). In this equation the observed electric field of a moving charge is contracted in all directions by the factor γ and additionally in the x direction by a further factor γ relative to the y and the z dimensions.

It will be shown that it is the combination of the relative velocity of the wire combined with the velocity of the current which cause the so-called 'magnetic' effect. But there are two separate and distinct causes to this effect dependent on whether the current and relative velocities combine either in line or at 90deg. to each other.

The Single Charge Equivalence

The total force on the unit test charge Q in Fig.(9.1) is

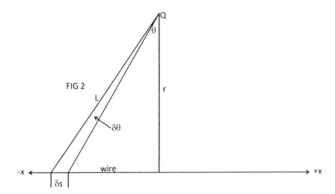

Figure 9.1: The Single Charge Equivalence

the integration of the separate forces from every individual charge - electron and ion - within the wire, as far as infinity in both directions. The total force from the positive ions and the negative current electrons are calculated separately and then subtracted from each other to give the net force on the test charge.

Rather than considering individual charged particles we may instead employ an arbitrary unit linear charge density, q_L along the wire.

In Fig. (9.1) the wire segment $ds = d\theta L/cos\theta$ as a consequence of the angle θ of the wire to the direction of the test charge Q.

The component of the force dF on the unit test charge in the L direction due to the segment ds obeys the inverse square law rule and is given by:-

$$dF = kq_L cos\theta ds/L^2$$

where $k = 1/4\pi e_0$.
Substituting for ds gives:-

$$dF = kq_L d\theta/L$$

Now $L = r/cos\theta$ thus:-

$$dF = kq_L cos\theta.d\theta/r$$

Integrating between $\theta = 0$ and 90deg gives

$$F = kq_L/r$$

Therefore the total force in the r direction from the wire on both sides of the test charge is:-

$$F = 2kq_L/r \quad\quad\quad (9.1)$$

Now this happens to be the equation for the electric potential field of a single charge of magnitude $2q_L$ positioned at the wire at $x = 0$.

Thus in subsequent calculations, to obtain the effect of relative velocity by all the like charges in an infinitely long straight wire upon the test charge Q we need only consider velocity effects on the electric potential field of a single theoretical charge of $2q_L$ positioned at $x = z = 0$.
As the test charge Q is a unit charge the force upon that charge is identical to the electric field E at Q which, in turn, is equal to the electric potential field at that point. Thus:-

$$F = E = \phi_{qL} = 2kq_L/r$$

The Wire inline with IRF Velocity
In the following calculations velocities are given as fractions of the speed of light.

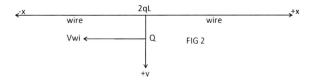

Figure 9.2: Charge Velocity Parallel to Wire

In Fig (9.2) the wire is taken to lie in the direction of relative movement (the x direction). The origin of all three axes is at the observer test charge.

The electric field of the current electrons, as seen by the observer, is a function of their velocity V_{oi} through the observer's IRF. Now V_{oi} is the relativistic addition of the wire velocity, V_{ow}, (with respect to the observer's IRF) and the current velocity, V_{wi}, (with respect to the wire). These three velocities create three different Lorenz Functions, γ_{oi}, γ_{wi} and γ_{ow}, to be employed in the observed electric potential field equation. These three functions can be inter-related via the Lorenz Velocity Transform equation, eqn. (13.11), as follows:-

$$\gamma_{oi} = \gamma_{ow}\gamma_{wi}(1 + V_{wi}V_{ow}) \tag{9.2}$$

where V_{wi} must be entirely in the x direction, ie. the direction of wire motion through the IRF - which is indeed the case.

Applying the single charge equivalence, q_L, positioned at $x = z = 0$ and employing the observed potential field eqn.(6.7) gives a force field at Q of:-

$$F_y = 2kq_L/\gamma y$$

where $k = 1/4\pi e_0$.

Subtracting the field of the current electrons from that of the ions at the position of the observer we obtain:-

$$F_y = 2kq_L/y\gamma_{oi} - 2kq_L/y\gamma_{ow}$$

As the velocity of the current electrons is very small relative to c (of the order of one millimeter per second) then γ_{wi} can be taken to be unity. Employing the conversion eqn. (9.2), gives the net potential at the observer of:-

$$F_y = -2kq_L V_{wi}(V_{wi}/2 + V_{ow})/\gamma_{ow}y$$

As V_{ow} is usually a small fraction of c and V_{wi} is small relative to V_{ow} we have:-

$$F_y = 2kq_L V_{ow} V_{wi}/y \qquad (9.3)$$

The net electric force F_y arises from all the current electrons and all the ions in the wire. It acts upon a unit observer charge in the direction normal to the relative velocity of the test charge inline with the wire.

It can be seen that there must be both a current through the wire and a relative velocity between the observer charge and the wire in order to produce the 'magnetic' force.

The Wire at 90deg to Relative Velocity
The effect which occurs when the wire lies in the direction of

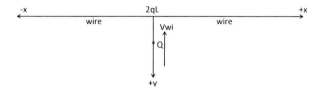

Figure 9.3: Charge Velocity normal to Wire

relative velocity no longer occurs in this case as the current electron velocity has no component in line with the velocity of the wire. The wire lies along the x axis but moves in the y direction relative to the observer.

Once again the equivalent theoretical single charge q_L is employed positioned at $x = z = 0$ and distance y from the observer charge Q.
The relative velocity (V_{ow}) effect (wire to observer charge) combines with the velocity (V_{iw}) effect (current velocity) on the electric field of the current electrons but now at right angles to each other.
The effect is to produce an elliptical shape $(z = 0)$ of the observed (by Q) electric potential field of the current electrons. The ellipse is rotated relative to the x and the y axes through the angle θ, where $tan\theta = V_{iw}/V_{ow}$ and is contracted in the direction of motion by the factor $(1 - V_{iw}^2 - V_{ow}^2)$.

For velocities very much smaller than c we have:-

$$\phi_y = 2kq_L/y \tag{9.4}$$

Now a tangent to the ellipse of the electric field at the position of the observer Q is not normal to the y axis due to the rotation of the ellipse described above. Hence there is a component of the electric field which produces a force in the x direction.

The tangent of the angle ψ to the ellipse at $x = 0$ is given by:-

$$tan\psi = (tan\theta - tan\sigma)/(1 + tan\theta.tan\sigma)$$

where σ is the angle of the ellipse to its major axis.

The tangent to a non-rotated ellipse is $tan\sigma = tan\theta.b^2/a^2$ where $a^2 = 1$ and $b^2 = (1 - V_{iw}^2 - V_{ow}^2)$.

Taking both V_{iw}^2 and V_{ow}^2 to be small relative to c we obtain the x direction component fraction to be:-

$$tan\psi = V_{iw}V_{ow} \tag{9.5}$$

Multiplying eqn. (9.4) by eqn. (9.5), gives the component of the charge acting in the x direction, normal to the relative velocity of the wire and the observer charge. There is no component of the field of the wire ions in the x direction for their field is not rotated as they are affected by only a single velocity.
Therefore:-

$$F_x = 2kq_L V_{iw}V_{ow}/y$$

It can be seen that the force F_x at right angles to the movement of the charged observer relative to the current carrying wire is identical for both inline or orthogonal movement and therefore is entirely independent of the direction of relative movement between the test charge and the plane of the wire. Thus:-

$$F_n = 2kq_L V_{iw} V_{ow}/r$$

where r is the normal distance from the wire to the observer and F_n is the force normal to the velocity of the test charge.

A positive sign for the force F_n indicates that the force is in one direction normal to relative velocity where-as a negative sign indicates that the force is in the opposite direction. The velocities V_{iw} and V_{ow} are signed for an arbitrarily chosen direction. It can be seen that changing the direction of either V_{iw} or V_{ow} reverses the direction of the force on Q.

Substituting the current I for $q_L V_{iw}$, replacing the factor k by $1/4\pi e_0$, renaming V_{ow} as V and returning to normal velocities gives:-

$$F = IV/2\pi e_0 c^2 r \tag{9.6}$$

which is the recognised force on a unit test charge where the relative velocity in the plane of the wire between the unit test charge and the IRF of the wire is V.

Converting to the terminology of magnetics where $\mu_0 e_0 = 1/c^2$ we have:-

$$F = VI\mu_0/2\pi r$$

which is the accepted formula for the magnetic effect.

So from the initial assumption that magnetism is not a fundamental force of nature these 'magnetic' phenomena are still derived.

It can be seen that the 'magnetic' - or perhaps the more properly named Relativistic Electric Effect - is caused by two different effects arising from separate relativity aspects of the electric potential field.

Applications of the Relativistic Current Effect
The Force between Separated Parallel Conductors.

Figure 9.4: The Force between Parallel Wires

The separation distance between the two conductors is d and their length is initially considered to be infinite.

In this exercise the observer test charge in wire B is the arbitrary linear unit charge density q_L of the current electrons times the wire segment ds.

From eqn.(9.6) the force exerted by wire A with current I_A on the charge $q_L ds$ in wire B is:-

$$F = I_A V_{iB} q_L ds / 2\pi e_0 c^2 d \qquad (9.7)$$

where V_{iB} is the velocity of the current electrons in wire B.

Substituting for $I_B = V_{iB} q_L$ and integrating over the length S of the parallel wires gives:-

$$F = I_A I_B S / 2\pi e_0 c^2 d$$

The force per meter in magnetic terms is:-

$$F = I_A I_B \mu_0 / 2\pi d$$

which is the standard magnetism equation for the force between two parallel conductors.

If the current velocities are in the same direction the force operates to move the wires away from each other and if the current velocities are in opposite directions the two wires are drawn together.

Induction as a Relativistic Current Effect

Inductance, symbol L, is the factor in an electric circuit which acts against any change in the current in that circuit. The changing current in the inductance creates a voltage, called the back emf (electro-motive force), which opposes the applied voltage causing the change of current. The back emf is found to be the product of the value of the inductance times the rate of change of the current through the inductance, thus $emf = -LdI/dt$. It is therefore impossible for the current to change from one level to a different level in zero time for this would be an infinite rate of change and would require an infinite voltage driver to overcome the induced back emf.

Inductance is inherent in all electric circuits to one degree or another. Even a single straight wire exhibits inductance. Inductance is currently considered to be a magnetic phenomenon.

A simple example of an inductance consists of two straight parallel wires of radius r length l separated by distance d and carrying the same current, but in opposite directions as shown in fig.(9.4).

The electric force F generated by the current I in wire A and experienced by the current electrons of wire B is given by the eqn.(9.7)

$$F = 2kIv_oq_Lds/c^2x$$

where v_o is the velocity of the electrons in both wires, $k =$

$1/4\pi e_0$ and x is the axis in line with the separation distance d.

Therefore the electric force E generated by wire A is given by:-

$$E = 2kIv_0/c^2x \qquad (9.8)$$

The electric potential difference from wire B to the surface of wire A is given by the integral of eqn.(9.8) between the limits of d and r therefore:-

$$p.d. = 2kIv_o/c^2 \int_r^d (1/x)dx = 2kIv_oln(d/r)/c^2$$

The potential difference between adjacent elements length ds of wire A due to a current gradient along the wire of dI/ds is given by:-

$$emf = -2(dI/ds)kv_oln(d/r)ds/c^2$$

putting $ds = v_odt$ in dI/ds and deriving the back emf per meter gives:-

$$emf/m = -2(dI/dt)ln(d/r)k/c^2$$

Now $emf = -LdI/dt$ so $L = 2ln(d/r)k/c^2$.

As the inductance is the same for both wires the above value is multiplied by 2 for this particular arrangement. Thus the total inductance is given as:-

$$L = ln(d/r)/\pi e_0c^2 = \mu ln(d/r)/\pi$$

The back emf is equally induced in any secondary wires parallel to the primary wire carrying the changing current as the induced potential field in the primary wire extends out to infinity. If the two wires are very close to each other then the back emf will be almost identical in both primary and secondary wires. When the secondary wire is connected to an electrical load then a current is induced to flow through the secondary but in the opposite direction to that in the primary wire, such that the two currents generate opposing back emfs.

All inductances of whatever physical configuration or geometry are considered to be subject to the same analysis.

The Conclusion

It can be seen that magnetic effects are explained as a consequence of the small velocity difference between the electrons and the ions within a wire which slightly modifies their electric fields so as to give a small net electric effect. This effect is only experienced by a charge with relative velocity with respect to the wire.

So why have physicist believed in the existence of a separate magnetic force for so long. The reason is surely because the force generated by a magnet appears so different from that generated by a charged body that the initial belief that the two forces were fundamentally different could not be entirely removed from physicist's minds, despite the halfway efforts of Maxwell to combine the two forces in his eponymous equations. But when assumed from the very beginning that the magnetic force does not exist then subsequent physics does not have a need for it.

When it is initially assumed that a magnetic force does exist then the subsequent physics developed from that basis always retains some aspect of the initial assumption, albeit the electric and the magnetic forces tend to merge together as per Maxwell's equations.

Modern Physics considers that radiated energy consists of electromagnetic waves having both an electric and a magnetic component at right angles to each other and of equal significance in the wave. It is now obvious that Modern Physics is wrong, for the waves can only be electric in nature. It was always difficult to understand how radiated energy could pass through a narrow slit, as either the electric or the magnetic component of the wave must be impeded at certain orientations. The consequence being that the structural integrity of the electromagnetic wave would be destroyed. On the other hand a purely electric wave would merely be diminished in energy and/or modified in polarisation - but its essential integrity would remain.

Electric waves are fluctuations in the electric potential of the Aether which fall off in amplitude with inverse of distance from their source, in exactly the same way that an elevated constant electric potential diminishes with distance from a charge source.

It is the gradient of the electric potential which accelerates charges. In the case of an electric sinusoidal wave the gradient is merely a phase shift of 90 degs to the electric potential.

Chapter 10

Electric Wave Transmission

In Modern Physics the transmission, reflection, refraction and diffraction of electric waves is explained by the theories of Augustin-Jean Fresnell and Gustav Kirchoff. In turn their theories were based to a large extent upon a theory by Christiaan Huygens. Thus, in this book the Modern Physics theory of wave transmission is referred to as the FKH theory.

FKH theory postulates that every point of a plane wave front advancing through Space acts as a separate wavelet generator which transmits the wave front *forwards* in all directions equally at the speed of light with an intensity which diminishes with inverse distance. The amplitude of these separate wavelets superposition upon each other at every point forward of the main wave front causing constructive or destructive interference at every point. Thereby a new advanced wave front is created. FKH theory is therefore based upon the assumption that each 'point' in Space acts as a separate source and recipient of waves.

It can immediately be seen that the FKH wavelet mechanism is almost identical to the Aether hypothesis that points in Space (Aethons) spread their individual influence in all directions by means of the Aether Static Transmission Mechanism. The FKH theory of wave transmission is clearly an Aether Theory, in particular because it recognizes the physical properties of points in Space.

The Aether Transient Transmission Mechanism
The Aether Static Transmission Mechanism describes how Aethons influence the electric potential of each other under substantially static conditions. But a dynamic version of the ASTM is required to deal with the movement of electric fields.

Across the Aether the Aether pressure (electric potential) varies as a consequence of the presence of matter and energy (waves). It is to be expected that these variations in pressure cause the movement of Aethons within the Aether matrix, which in turn causes changes in local Aethon density. However the degree of Aethon movement is possibly very small and the Aether matrix may be merely distorted rather than corrupted. The movement of Aethons, however minute, requires a change in velocity and hence an acceleration of the Aethons. As pressure is required to accelerate Aethons then it follows that they must possess inertia. In all these aspects the Aether acts in a very similar manner to a crystal constructed of identical atoms through which acoustic waves may propagate. However, pressure, inertia and velocity with respect to the Aether may be rather dissimilar to those same parameters when applied to matter.

The original impetus of Aether movement must initially be derived from a source of some kind - such as an oscillating charge. The waves or transient then move away from the originating source at the local propagation velocity of the

Aether and under the action of the inertia of the Aethons. The direction away from the originating source is called 'forwards'.

Every Aethon within the wave or transient may be considered to be a 'source', as it passes its electric potential (pressure) difference on to the contiguous Aethons immediately in front. However an Aethon is not a true wave source but merely a stepping stone. Each Aethon receives signal (time-varying electric potential) from every Aethon in all directions behind the plane normal to the signal direction which it occupies, delayed by the time d/c and diminished with inverse distance. This multiplicity of signals are all super-positioned (combined) at each receiving Aethon. The resultant signal is then passed on to every Aethon in all directions forwards of the plane. Thus every Aethon simultaneously acts both as a collector and as a transmitter of signal.

In the above arguments distance in all three directions is determined by the Aether matrix.

The Advancing Wave

Take the case of a flat plane wave-front shown in Fig.(10.1) - but temporally ignoring the screen - of infinite dimensions and advancing in a direction normal to its surface where the electric potential at the wave-front is oscillating at the identical frequency and is everywhere in phase across the wave-front. According to both Aether and FKH theory the electric field at point P arises from the constructive and destructive interference contribution at that point from every Aethon of the advancing wave-front. The contribution from a nominal Aethon at the point D on the current wave-front is diminished by the factor $1/r$ and phase shifted by the factor $\omega r/c$ due to the time to transit distance r.

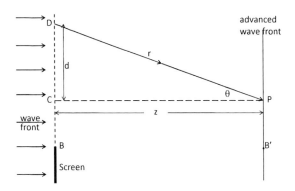

Figure 10.1: An Advancing Wavefront

Aethons situated in the wave-front at large distances from point C (distance d is large and thus r is much larger than z) have little effect upon the potential at P due to the inverse distance effect upon magnitude combined with a rapidly varying phase shift giving a net zero signal.

The mathematical calculation of the resultant potential at point P is complex but is standard textbook and so is not fully explained here. It is found that the effect at point P is identical for every point situated at the same distance z from the current wave-front. The potential at P is exactly the same magnitude and frequency as that in the current wave front except that it is phase shifted by the factor $\omega z/c$. Thus the original wave-front is advanced unchanged, except in its phase, at the speed of light.

Cornu's Spiral
Out of interest, the amplitude of the wave at point P can be obtained from the application of Cornu's spiral. Cornu's spiral is obtained by superpositioning small arrows, the length

Figure 10.2: Cornu's Spiral

of which represent the magnitude of signal received at point
P from a unit area of the advancing wave-front of length
δd and width δy at distance d from point C. The direction
of the arrow represents the phase shift of the signal, which
increases with the square of distance d. The tail of each
arrow is attached to the point of the previous arrow. The
diagram of an infinite number of arrows from a wavefront
stretching to infinity in both directions from point C gives
Cornu's spiral. The amplitude of the signal at point P for
an infinite wave front is obtained from the distance between
the two ends of the spiral.

Edge Diffraction
However, the conditions at the boundary or edge of a flat
plane wave-front are not transmitted forwards in an exact
repetition of the original wave-front. Consider that point
P is positioned either side of point B' opposite the screen
edge. The super-positioned electric potential on the advanced wave-front near to point B' no longer arises equally
from the wave front either side of point B, as the screen
blocks out the wave-front below point B. The effect of
the missing contribution of signal is to truncate one end
of Cornu's spiral and thus modify the amplitude of the received signal at and within a few wavelengths either side of

point B'.

Consider that the advancing wavefront is a parallel light beam partially obscured by the screen. We would intuitively expect that beyond the area of the wavefront not obscured by the screen the strength or amplitude of the light beam would be constant throughout. Immediately behind the screen we would expect to find total darkness. In fact the truncation of Cornu's spiral by the screen causes the light amplitude from the screen edge in the no screen direction to oscillate in amplitude to a diminishing degree with distance from the screen edge. In the opposite direction from the edge and behind the screen there is experienced a degree of light, totally unexpected, but which falls off rapidly in magnitude within a few wavelengths of the edge. These effects are called diffraction.

Dispersion from a Small Hole

The boundaries of a wave-front, eg. a light beam, may be

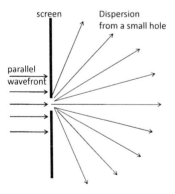

Figure 10.3: Small Hole Dispersion

determined by a screen with a hole of a certain dimension

placed in the path of a wider wave-front such that the emerging wave-front is of the dimensions of the hole, Fig.(10.3).

The effects of diffraction are then observed around the edges of the emerging beam at some distance from the hole. But when the hole diameter is made small relative to the wavelength of the signal the hole then acts rather like a single Aethon source, passing the electric signal forwards in all directions in a hemi-spherical wave front. In fact, light is seen at angles to the original path where no light is seen when the hole is bigger. This effect occurs as the paths to any point forward from opposite sides of the hole differ very little in length due to the small dimension of the hole. This gives minimum phase-shift with the consequence that neither constructive nor destructive interference occurs in the advancing wave-front. With the bigger hole destructive interference removes the signal at extreme angles of deflection.

The phenomenon of diffraction allows electric waves, such as those of light, to bend round objects at considerable angles and take an entirely different course to the original wave path.

Diffraction is strong evidence for the existence of the Aether.

The Least Time Theorem

In advancing from one point in Space to another a wave always takes the path of least time between the two points. This is the theory of P. Fermat. It is immaterial whether the path is direct, reflected by a mirror or refracted by a lens.

It would appear that Fermat's theory implies that a light ray previously knows which path is the one of least time before it takes that path. But this is a misunderstanding of what really happens. Instead, the light ray actually takes an infinity of different paths. Most of these paths will be longer than the path of least time and consequently will be of a different duration. Rapidly varying length paths give rapidly varying phase shifts and a mix of many different phases equally constructively and destructively combine at the final destination. The net result is a zero signal.

A signal only results when many paths have identical, or almost identical, duration and hence identical phase shift. This situation only occurs where the paths are situated close to and around the path of least time. Thus it is the path of least time which determines whether a wave is reconstructed or not.

Hence the wave is the servant of the path.

Chapter 11

Gravity

Newton's Gravity

Our everyday experience tells us that material bodies are attracted towards the Earth and, if that attraction is not inhibited in any way, they fall towards the Earth. That action we call gravity.

We observe that different types of body fall at different velocities. A feather, for instance, falls at a very much slower rate than a stone. This is because, when bodies fall through the Earth's atmosphere, they are subject to two different forces. On the one hand they are subject to the force of gravity in the direction of Earth and they are also subject to a friction force in the opposite direction which arises from contact with molecules of the air as the body moves downwards. As the friction force is a function of the velocity of the body the two forces cancel each other at a particular velocity with the result that eventually there is no further acceleration of the body. The object continues to fall but at the same constant velocity, called the terminal velocity. Terminal velocities depend greatly on the density of the falling object which is why stone falls faster in air than a feather.

However when objects fall through Space there is no atmosphere to cause friction and so there is only one force in operation - the force of gravity. It was Galileo who first discovered that gravity - unlike the electric force - accelerates all material bodies, of whatever substance, at exactly the same rate. In the emptiness of Space it is found that a feather falls at exactly the same rate of acceleration as a piece of lead. This fact is known as the Weak Principle of Equivalence.

Sir Isaac Newton was the first to realize that the action of gravity extended beyond the atmosphere of our planet into Space to a great distance. For example, it is the gravitational field of the Earth acting upon the Moon, a quarter of a million miles away, which keeps the Moon in its orbit around the Earth and so stops the Moon from flying off into Space. On the other hand it is the gravitational field of the Moon that causes the tides here on Earth.

Newton also realized that every massive body, however large or small, generates its own gravitational field.

Because the force of gravity does not appear to be experienced inside a space-ship in orbit around the Earth most non-scientific people believe that the force of gravity does not exist in Space. This is entirely wrong as Newton showed. The orbiting space-ship (rockets not firing) is in fact continuously accelerating towards Earth under the influence of Earth's gravitational field. As the astronauts within a space-ship are subject to exactly the same gravitational acceleration as the space-ship there is no relative force to pin the astronauts to the floor or the walls of the space-ship - and so they just float around inside the cabin.

The significant factor required to maintain an orbit is a degree of velocity in a direction normal to the gravitational acceleration. The space-ship continuously falls towards the centre of the Earth but at the same time its velocity takes the space-ship *around* the Earth with the effect that the space-ship never gets any closer to the Earth.

All orbits are actually elliptical in shape. The degree of normal velocity determines the degree of ellipticity of the orbit. A very specific velocity is required in order to give a perfectly circular orbit.

From a study of the orbits of the planets as they move under the influence of the Sun's gravitational field Newton calculated that the degree of acceleration of the field diminished with the inverse square of the distance from the centre of the massive body causing the field. Newton was then able to formulate his famous equation stating the Law of Gravity in terms of gravitational force:-

$$F = GMm/r^2$$

where M is the mass of the source body, m the mass of an object body at distance r from the centre of the source body and G is a fundamental constant, the Gravitational Constant.

Although Newton's equation is extremely well known in the above form, strictly speaking it is not the purest form of the gravitational equation, for gravity generates an acceleration rather than a force. As force equals mass times acceleration ($F = ma$) Newton's formula may be modified to the more philosophically correct form:-

$$a = GM/r^2$$

What we call forces are really the result of opposing accelerating fields. For example, as we stand upon the surface of the Earth we feel a force on the soles of our feet. This force is due to the acceleration of gravity towards the centre of the Earth in opposition to the acceleration upwards from the electric fields generated by the atoms of the Earth upon the soles of our feet.

The Cause of Gravity

Gravity is simply the effect of refraction in and by Space.

Refraction is an effect caused by differing velocities of light, which may be evidenced in everyday objects such as lenses. Every spectacle wearer employs the effect of refaction to enable themselves to see more clearly at a great distance - or perhaps to read print at a close distance. A lens works because the effective speed of light through glass is less than that through air. Light passing through the middle of the lens where the glass is thickest is slowed more than light passing through the thinner outer regions of the lens. The time delay through each part of the lens is arranged such that the total travel time, both through glass and air to a particular point beyond the lens, is exactly equal. That point is the focal point of the lens.

With a glass lens light either moves at one velocity through the glass and at a different velocity through air. However, in Space the speed of light varies in a continuous manner from point to point. This is a surprising statement to make for we are also told that the speed of light is a universal constant.

It is the case that the difference in the speed of light between two points in Space is the gravitational potential between those two points - hence it is the presence of mass which causes light to slow. For example, at the surface of the Sun the speed of light is less than that at a great distance by the factor 4.2×10^{-6}.

It can be seen that the difference is very small but even so it causes a gravity field 28 times as strong as that on the surface of the Earth. So, except in the vicinity of exotic objects such as neutron stars, where the mass density is many, many times greater than that of a normal star, the speed of light is *nearly* constant throughout the Universe. However the difference, albeit very small, is the cause of the all important gravitational force.

Why the Speed of Light Slows

It was postulated in the chapter on the electric field that the propagation velocity of the Aether, the speed of light, is given by:-

$$c^2 = C_e/\rho_e$$

The speed of light is a function of both mass-less pressure (electric potential) per unit deformation, C_e, and Aethon density ρ_e.

Now a gravitational field may exist where there is zero electric potential (ambient Aether pressure), therefore the observed slowing of light close to an electrically neutral mass indicates that the effect must be caused by increased Aethon density alone. In turn, the increased Aethon density must be caused by the presence of the gravitational source, the massive body. Now an increase in Aether density can only be obtained from a movement of Aether towards that point. But if mass merely displaced Aether from within the body of the mass to outside of the body - increasing density outside and diminishing it inside - then the increased density outside would eventually dissipate into the far regions of Space leaving no excess of density local to the mass. Therefore in order to maintain a permanent local increase in Aether density close to mass it is postulated that:-

Mass creates new Aether at a rate in proportion to the amount.

The effect of this new Aether is to increase the Aethon density both within and immediately outside of the source mass. The increased Aethon density then diffuses outwards via Aethon to Aethon contact to an ever decreasing level.

This generation of new Aether has two significant cosmological effects. Firstly the gradient of the decreasing Aether density with distance causes the phenomenon of gravity, whilst the additional Aether contributes towards Cosmic Expansion. Thus gravity and cosmic expansion are different effects of the same mechanism.

As the average density of mass in the Universe is very small it would appear that new Aether generated by mass alone might only account for a small fraction of Universal cosmic expansion. Thus there is likely to be other causes of new Aether generation. This is discussed in the chapter on Cosmic Expansion.

The presence of a matter body increases the immediate Aethon density by an unknown amount ρ_e. Thus:-

$$c_L^2 = C_e/(\rho_\infty + \rho_e)$$

where c_L is the local speed of light.

Now employing an identical Aether shell to Aether shell diffusion process over increasing distance from the mass body to that employed in the diffusion of the electric potential we obtain, $\rho_e = \rho_1/r$ where ρ_1 is the increased density at an arbitrary distance of one meter from the mass, on the theoretical assumption that the entirety of the mass is situated at a single point. Thus:-

$$c_L^2 = C_e/(\rho_\infty + \rho_1/r) = c_\infty^2/(1 + \rho_1/\rho_\infty r)$$

It can be seen that no matter how large the mass induced additional Aether density the speed of light can never diminish to zero.

Alternatively we may have:-

$$c_L = c_\infty(1 - \rho_1/2r\rho_\infty)$$

when the ratio $\rho_1/r\rho_\infty$ is very much less than unity, as it is at the suface of most celestial bodies.

The datum point for c_∞ is taken to be at infinite distance from the source mass but may be arbitrarily positioned. In future the datum speed of light is simply given as c.

We now take $\rho_1/2\rho_\infty$ to equal Am where m is the gravitational radius of the mass ($m = GM/c^2$) and A is as yet an unknown factor, but is later determined to be 2.
This can be said as the ratio ρ_1/ρ_∞ is determined by the postulate to be a function of mass M. The gravitational constant G incorporates the other factors.
We now have:-

$$c_L = c(1 - Am/r) \qquad (11.1)$$

or alternatively

$$c_L - c = \Phi = -Amc/r \qquad (11.2)$$

Thus the gravitational potential $c_L - c$ diminshes with inverse distance from the source mass. Hence the gradient of the potential difference diminishes with the inverse square of distance in agreement with Newton's law.
This equation leads to almost identical predictions on the behaviour of both light and mass passing through or within a gravitational field as the General Theory of Relativity (GR). The essential difference between the two theories lies in their different explanations of the physical representation of the gravitational field and in the mechanism of the cause of that

field.

The Gravitational Field

The physical gravitational field is now seen to be a field of gravitational potential Φ rather than a directly accelerating field. Thus:-

$$\Phi = -AGM/r \qquad (11.3)$$

where the factor A is the conversion factor between the acceleration of light versus mass in the same gravitational gradient.

From eqn.(11.3) it can be seen that the gravitational field extends to infinity in all directions, diminishing in magnitude with the inverse of the distance - although this does not remain true at great distances and weak fields where a different factor comes into play.

The mechanism of the diffusion of excess density via Aethon to Aethon contact intrinsically determines that the gravitational fields of multiple sources combine at each and every point in Space to create the ambient gravitational field. The accelerating ability of the ambient field is a function of the gradient of the field at each point in Space. Just as with the electric potential field it is the local ambient gravitational field gradient which directly acts upon an object mass, rather than the individual sources of that field. The nearest and strongest source is naturally more significant in its effect upon the field than the further and weaker sources. Although it might appear that a particularly significant mass is directly causing the acceleration of an object body, nevertheless the effect is only indirectly connected to the mass via its contribution to the ambient field.

The gravitational acceleration of matter is expected to occur in a similar manner to the electric field in that a speed of light gradient unbalances and so modifies the internal asymmetric geometry of an FMP such that the Aether velocity of the FMP is changed.

Non-Euclidean Space

A variation in the density of the Aether means that Space is not Euclidean. One example of non-Euclidean Space is that the angles of a triangle drawn in that Space no longer add to 180 deg. A further example is that the circumference of a circle no longer equals 2π times its radius.

Non-Euclidean Space simply means that the 'density' of Space is not uniform throughout. However a variation in density requires a more fundamental standard of distance against which density is determined. Hence there must also exist a sub-Aether more fundamental than the Aether. If this seems to be generating a rather too complex picture of the Universe it should be remembered that General Relativity also requires two levels of distance standard. The sub-Aether units of distance are equivalent to the GR coordinate units while the distance units for the Aether are equivalent to the GR proper or cosmological units.

The density of the Aether is determined as the number of Aethons, ie. Aether volume per unit volume of the sub-Aether.

In an area of the Aether where the Aethon density is relatively high the Aethons are closer together. In the Aethon world the unit of distance is defined as an Aethon - on the other hand the Aethon separation distance as determined by the sub-Aether is a variable. Thus Aether density cannot vary when viewed from the Aether but when viewed from the sub-Aether it does vary. Matter and energy exist in the Aether Universe and so cannot detect the Aether density but the variation and the gradient of Aether density does have a significant effect, the effect we call gravity.

The atmosphere of the Earth is a common example of a substance with variations in density. Not only is the air more dense at low than at high altitude but the density varies with local pressure in the weather conditions of cyclones and anti-cyclones. One consequence of a non-Euclidean atmosphere is that sound, and to a much lesser extent light, no longer travel in straight lines. For example the light from the setting Sun travels tangentially to the Earth's surface across the air density gradient. As light travels slightly slower through denser air there exists a speed of light gradient across the beam such that the sunbeams are bent to a small degree towards the Earth.

Gravitational Redshift

An immediate effect of the speeding up of a light ray as it recedes from a source of gravity is an increase in the wavelength of the ray. Thus a visible light ray is shifted towards the red end of the spectrum (a redshift). Conversely, photons are blue-shifted in approaching a gravitational source. Thus the observed frequency of the ray at distance from the source mass is observed to be lower than its frequency at the point of generation. Giving:-

$$f = f_\infty(1 - 2m/r)$$

The gravitational red-shift effect was conclusively detected by Pound and Rebka in 1959 at the Harvard University in the minute gravitational potential difference existing over the height of the Jefferson Tower. They were able to measure the very small redshift obtained by employing the extremely stable frequency of the Mossbauer effect. This experiment (and others) determines the value of A in eqn.(11.1) to be 2.

As an example of the gravitational redshift effect con-

sider two identical atoms, both emitting light at a frequency
characteristic of that particular atom. One atom is on Earth
and the other is stationary on the surface of the Sun. The
light from both atoms is viewed from Earth. The frequency
of the light from the solar atom decreases in leaving the Sun
and arriving at Earth so the Earth based observer will consider that the solar atom emits 'pulses' more slowly than
the identical Earth atom. His conclusion will be that time
passes more slowly on the surface of the Sun than on the
surface of Earth.

The Bending of a Light Ray crossing a Speed of Light Gradient

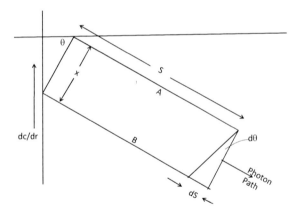

Figure 11.1: Refraction

Fig. 11.1 describes a photon of unknown width x crossing
a speed of light gradient at an angle θ to the gradient.
Due to the speed of light gradient the upper side A of the
photon moves a greater distance in an equal time dt to the

lower side B by an amount dS. Thus:-

$$d\theta = dS/x = \frac{dc_L}{dr}x sin\theta dt/x = \frac{dc_L}{dr}sin\theta dt \qquad (11.4)$$

The bending effect $d\theta/dt$ is an example of refraction.

As acceleration equals $vd\theta/dt$ generally, then the acceleration of the light ray in a direction at right angles to its path is given by:-

$$acc = csin\theta\frac{dc_L}{dr} \qquad (11.5)$$

The acceleration down the speed of light gradient is given by:-

$$acc_r = csin\theta^2\frac{dc_L}{dr} \qquad (11.6)$$

When the light ray path is normal to the speed of light gradient $(sin\theta = 1)$ the acceleration is just cdc_L/dr.

Differentiating eqn.(11.1) gives:-

$$acc_r = cdc_L/dr = Amc^2/r^2 = AGM/r^2 \qquad (11.7)$$

As the acceleration of mass in a gravitational field is GM/r^2 it can be seen that light accelerates at A times the rate of mass at the same point in a gravitational field.

Tests of the Theory
Appendix 1 calculates the degree of bending of a light beam passing across a speed of light gradient as described by eqn. (11.4). In the example given the gravitational field employed is at the surface of the Sun where the field is 28 times as strong as that at the surface of the Earth. The light beam

from a distant star passes tangentially just above the surface of the Sun and is observed on Earth.

The gravitational bending of a light ray was first observed by A. S. Eddington who travelled to the island of Príncipe near Africa to watch the solar eclipse of 29 May 1919. During the eclipse, he took photographs of stars in the region close to the Sun and measured their position relative to stars further away from the Sun. The apparent relative position of near and distant stars was then compared to their positions when the Sun was in a different part of the sky such that its gravitational field did not seriously bend the light rays. This bending effect is noticeable only during an eclipse, since otherwise the Sun's brightness obscures the stars. Eddington's observations, and subsequent more accurate ones, observed the deflection to be 1.75 arcsecs which determines the value of factor A in eqn. (11.1) to be 2.

Although the acceleration of light is twice that of matter the effect on the transverse beam passing close to the Sun is very small. This because the light beam is only in the intense part of the Sun's gravitational field for a very small time. If it is taken that the intense part of the field stretches for about two million kilometers then the light beam is exposed for only 7 secs. Further more that component of the gravitational gradient which bends the ray diminishes rapidly at distance from the Sun. Thus the path of the beam is very little affected even in such a strong gravitational field.

The bending effect of refraction in the Sun's gravitational field may equally be considered to be an acceleration of the light beam towards the Sun. However when the beam is moving in an almost straight line we probably call the effect a bending. But consider if by some means we could restrain the path of the light ray into a circle with the plane of the circle normal to the speed of light gradient. Then the light ray would be bent continuously down the gradient with the resultant effect that the light ray circle would spiral down the speed of light gradient at an ever increasing rate. We might then prefer to say that the light ray circle fell in the Sun's gravitational field in a similar manner to the effect upon a matter body. But it would fall at twice the rate.

Appendix 2 demonstrates that Space is more dense close to a massive body. As light slows in passing through a gravitational field it follows that the time for a light ray/radio wave to reach a distant reflector and return to the sender will increase if a gravitational field is placed in the path of the ray where no field previously existed. The experimental method employed is to emit a radar beam from Earth which passes close to the surface of the Sun and is then reflected from a spacecraft or planet in a known position on the other side of the Sun. The reflected ray also passes close to the Sun on its journey back to Earth. The round trip time of the radar beam is accurately timed so as to give the distance to the reflector via the nominal speed of light. The position of the reflecting body must be very accurately predicted from previously known positions and velocities. The difference between the predicted distance to the reflector and the radar measured distance gives a measure of the increased density of Space due to the presence of the Sun's graviational field. This experiment is a far more accurate test of the variation in the speed of light than the measurement of the bending of a grazing light ray described in appendix 1.

An experiment of this type was first envisaged in 1964, by Irwin I. Shapiro. Shapiro proposed that radar beams be reflected off the surface of Venus or Mercury. When the Earth, Sun, and Venus were most favorably aligned, Shapiro predicted that the expected time delay difference between the presence and the absence of the Sun close to the signal

path of the radar signal would be about $200\mu secs$ - a measurement well within the limitations of 1960s era technology.
The first test, using the MIT Haystack radar antenna, was
successful, matching the predicted amount of time delay.
This experiment has been repeated many times since, with
ever increasing accuracy. This experiment also confirms that
the factor $A = 2$.

The gravitational effect of the slowing of a light ray may
be interpreted equally as a slowing of local time or as denser
Space as the gravitating mass is approached.

Appendix 3 describes how the orbits of the planets around
the Sun are affected by the distortion of Space around the
Sun. It is demonstrated that the gravitational field causes
a precession of the perihelion of each orbit. This precession
cannot, of course, be observed unless the orbit is an elliptical one. The orbit of Mercury is more elliptical than that
of any of the other inner planets of the solar system and it
had been noted by astronomers that the perihelion advances
a little each orbit. Calculations using Newtonian mechanics
predict a precession of 532 arc secs per century as a result of
the gravitational effect of the other planets, notably Jupiter.
However the observed precession is 42 arc secs greater than
this figure. Both GR and the Aether Theory of Gravity predict this additional degree of precession, or close to it.

The Effect of a Speed of Light Gradient on Mass
The Aether theory of gravity cannot fully explain how a
gravitational potential field causes an accelerating effect upon
matter bodies. In order to achieve that a complete knowledge of the internal construction of a fundamental matter
particle (FMP) is required - and we do not yet possess that
knowledge. The theory merely states that a matter body
accelerates at one half the rate of a transverse light ray at

the same point in a gravitational field - for that is what is observed.

The figure of one half is justified by observation but there is also some scientific logic to it.

In the first chapter it is proposed that an FMP is constructed of electric waves in some form of rotatory geometry. It follows that if the rotation of the electric waves is equally in all directions for an FMP stationary in the Aether then the rotatory waves will have components equally transverse to the speed of light gradient as in-line with the gradient. Thus the bending of the waves by the speed of light gradient occurs for only one half of the time. It should be remembered that according to the hypothesis of the construction of an FMP the bending of the internal electric waves permanently modifies the asymmetric geometry of the waves which causes the FMP to move through the Aether at a specific fraction of the speed of light. Thus an ambient speed of light gradient causes a change of velocity of the FMP in proportion to the magnitude of the gradient and the duration of exposure.

The factor of one half, $1/A$, applies only to an FMP with zero Aether velocity. At higher FMP velocities in the direction of the speed of light gradient the FMP geometry is increasingly distorted in its asymmetry. At a velocity approaching the speed of light the mass of a body approaches infinity, $m = \gamma m_0$, and the factor of one half the acceleration of light diminishes to near zero. It is intuitively understandable that the greater the geometric distortion the more difficult it is to produce further distortion from an identical stimulus. On the other hand, when a matter body crosses a gravitational gradient transversely at near the speed of light then the factor of one half may approach unity.

Inertia

Inertia is mainly a condition of Space, rather than of light
or matter, where the gradients of the electric and the gravitational potentials are absolutely zero. In the absence of
these gradients light propagates at a constant velocity along
a path, which we may call a straight line, determined by
the complete regularity of the Aether matrix. In zero gradient matter also moves in a straight line path at a constant
Aether velocity. In the case of matter the constancy of velocity depends upon the constancy of the internal geometry
of mass particles.

The absolute velocity and acceleration of a matter body
cannot be measured directly against the reference frame of
the Aether as the Aether cannot be directly detected. These
parameters can only be measured relative to an arbitrarily
chosen matter body, which itself may be moving and accelerating. However when the reference body is chosen to be a
distant star its own velocity, and hence transverse angular
movement as seen from Earth, becomes negligible as a consequence of the great distances involved. Thus distant stars
may reasonably act as markers for the Aether against which
local accelerations may be measured.

Ernst Mach suggested that inertia was determined by,
and therfore relative to, these celestial bodies. In fact they
merely mark the position of the Aether relative to which absolute velocity, and hence change of velocity, is determined.

The Pioneer Anomaly

The spacecraft Pioneer 10 and 11 were designed to pass right
out of the solar system. They are still receding from the Sun
and continuously decelerating under its gravitational attraction. Both spacecraft are now out of communication with
Earth at distances beyond 50 AUs. While they were still

in communication their position and velocity could be measured with great accuracy and compared to a theoretical position and value determined by the application of Newtonian gravity. It was found that the actual acceleration towards the Sun was greater than the theoretical figure by a factor of approximately $9 \times 10^{-10} m/s^2$. An exhaustive examination of all possible explanations failed to uncover the cause of the difference.

The explanation for the Pioneer anomaly would appear to lie in the effect of cosmic expansion coupled with the Aether theory of gravity. In the Aether theory the increased Aether density created by mass diffuses outwards from the source mass and diminishes in intensity as the shell area increases with the square of the distance. However cosmic expansion causes the volume of the Aether to expand over the time of transmission from the source of the increased Aether density to a point in Space distant from the source. Thus the volume of successive concentric Aether shells increases more rapidly than by the square of the distance. The result is that the gravitational potential falls off more rapidly than the inverse square of the distance, causing an increase in the inward gravitational acceleration above Newtonian levels.

Consider an Aether shell at radius r.
The elapsed time from the source over distance r at the speed of light is r/c so the cosmic expansion affect on distance is $(1 + H_0 r/c)$ and on the surface area of the shell it is $(1 + H_0 r/c)^2$.

The factor $H_0 r/c$ is generally very much smaller than unity thus the shell volume increases by:-

$$(1 + 2r H_0/c)$$

Therefore the density falls off more rapidly with the increased volume, thus:-

$$c_L^2 = C_e/\rho_e(1 - 2rH_0/c)$$

Therefore approximately:-

$$c_L = c(1 + rH_0/c)$$

where rH_0/c is a small fraction and H_0 is not considered to be a function of r.
The acceleration of a transverse light ray is given by:-

$$acc = cdc_L/dr = cH_0 \qquad (11.8)$$

For matter the acceleration is one half of that at $cH_0/2$.
Thus cosmic expansion creates an additional acceleration towards the source.

When H_0 is take to be $75km/s$ per mega-parsec the additional acceleration calculates to be $3.7 \times 10^{-10}m/s^2$ which is of the same order as the Pioneer observations, although only 0.41 of the observed value. This difference might be explained by cosmic expansion at the position of the Pioneer space craft being 2.4 time the Hubble value.

The Hubble value is, of course, an average expansion of the Universe measured over millions of light-years of Space and may well not reflect the local cosmic expansion within our solar system or within our galaxy.

The Galactic Orbital Velocity Anomaly

An observation of the orbital velocity rates of the outer stars of galaxies disclosed that the velocities were not diminishing with inverse distance from the galactic center as expected. In fact, stars at the galactic edge move at similar velocities to those halfway to the galactic center. This effect is caused by a higher than calculated level of gravitational acceleration but at very low levels than is predicted by Newton or Einstein. The current accepted explanation is that a halo of unseen and otherwise undetected 'dark' matter exists around the galaxy which creates an additional gravitational effect. The amount of dark matter required is possibly as much as ten times the amount of matter in all the stars of the galaxy. Being 'dark', ie. not emitting light, this new form of matter cannot be seen, but presumably should occlude the light from stars behind. However this effect is also not observed.

An alternative, but not widely accepted explanation for the constant orbital velocity, is called Modified Newtonian Dynamics (MOND). The theory simply states that Newtons law must be modified by some factor but gives no explanation as to how it occurs. The cosmic expansion effect within the Aether theory of gravity gives an explanation for the Pioneer anomaly (and hence MOND). It possibly also explains the observed galactic orbital velocity effect. However the local cosmic expansion would need to be less than the universal Hubble constant, at a level of only 24 km/s per mega-parse. Unfortunately the calculation of the rate of cosmic expansion within the various geometries of galaxies is beyond the knowledge and capability of the author.

But the currently accepted Dark Matter explanation requires an increase in the total mass of the Universe by a factor of ten in order to explain a gravitational anomaly of the order of just $1.2 \times 10^{-10} m/s^2$. That seems most unlikely.

Black Holes
In the Aether Theory of Gravity (ATG) the speed of light can never diminish to zero no matter how heavy or dense the source body. Thus all bodies may emit light. However,

if the emitted light does not travel exactly along the Aether density gradient it may well be bent completely round by the intense speed of light gradient which surrounds a highly dense matter body, such that the light returns to the source body. Thus the amount of light emitted by high density bodies may be greatly diminished, and as a consequence only a fraction of the light emitted will escape into outer Space.

I leave it to others to do the calculations to determine the fraction for a particular strength gravitational field if they wish. The Aether theory of gravity therefore does not rule out Grey Holes (as opposed to completely Black Holes).

General Relativity

Einstein's Theory of General Relativity (GR) is the theory of gravity currently accepted by the scientific establishment. GR and the ATG make identical predictions of the gravitational effect upon both light and matter but differ at very low levels of gravitational acceleration where the affect of Cosmic Expansion increases acceleration in the ATG relative to the predictions of GR. However the two theories give radically different explanations of the causes and the mechanism of gravity. For example GR considers that the action of gravity results from the warping of Space-Time but gives no explanation as to how matter causes the warping despite the indisputable fact that one must be the direct consequence of the other.

A further major difference between the two theories is that GR incorporates a time dimension into its postulate of Space-Time, where-as Aether theory denies the existance of a time dimension. The situation with regard to time is even more complex in that the form of time incorporated in GR theory is an imaginary time dimension rather than a real one, in that real time is multiplied by the imaginary factor $\sqrt{-1}$. As a consequence Space-Time cannot be envisaged in a physical form.

Furthermore GR gives no explanation as to cause of the exact determination of the local speed of light.

A further major difference lies in the different explanations of the two theories for the action of matter in a gravitational field. The ATG postulates that this action arises from an hypothesis for the construction of fundamental mass particles described in a previous chapter. The weakness of the ATG is that the internal structure of an FMP is not completely known so that the exact mechanism of acceleration cannot be fully described. On the other hand in order to accomodate matter GR employs one geometry of Space to explain the gravitational behaviour of mass described by:-

$$ds^2 = dx^2 + dy^2 + dz^2 + d\tau^2 \qquad (11.9)$$

where $\tau = \sqrt{-1}t$ and employs a different geometry to describe the behaviour of light through Space.

$$ds^2 = dx^2 + dy^2 + dz^2$$

It does not seem acceptable that Space can possess separate geometries to accomodate whichever particle happens to be moving through at the time.

Chapter 12

Space Expansion

Galactic Redshift (Z)

Hot bodies emit electric waves over a continuous band of frequencies - the hotter the body the higher the frequencies and the more intense the radiation. An iron bar heated to 500 degC will glow red - a colour at the lower end of the visible spectrum of frequencies - while a standard light bulb filament operating at about 3000 degC glows white-hot. The frequency spectrum of the light bulb is shifted upwards by the higher temperature into the visible spectrum which covers the range of colours from red to violet. The surface of the Sun is higher still at 5800 degK and the frequency range of emitted radiation extends well beyond the visible spectrum into the ultra-violet.

The important aspect of this solar radiation from the point of view of this chapter is that here on Earth we do not receive the whole spectrum of the Sun's emitted frequencies. Certain specific frequencies are absorbed by atoms in the Sun's atmosphere and so are not received here on Earth. An instrument known as a spectroscope has the ability to spread out the Sun's radiation spectrum so that the different frequencies can be seen as different colours in a manner rather similar to the way that water droplets in the atmosphere create a rainbow when the Sun shines upon them. The absorbed frequencies are then seen as fine black lines at various points in the spectrum. The frequency and/or wavelength of these absorption lines can be very accurately measured by the spectroscope.

As each different element absorbs radiation at its own specific frequency or frequencies the absorption lines enables us to determine which elements are present in the atmosphere of the Sun. In fact the element Helium was discovered in the Sun's atmosphere by this method before its presence was discovered here on Earth. Thus spectroscopy and the absorption lines enable us to discover that the Sun is predominately made of hydrogen with a certain amount of helium and a trace of many other elements.

It is equally possible to examine the light emitted by distant stars. Most stars are in fact very similar in construction to our Sun. The absorption frequencies of elements are identical for a particular element no matter where that element is in the Universe. Consequently specific elements can be detected in stars many millions of light years away.

There are two factors which shift the frequency of the entire emitted spectrum and hence the frequency of the absorption lines. Firstly, a decrease in gravitational potential at the surface of the source star lowers the frequency towards the red end of the visible spectrum - called a gravitational redshift. As the gravitational potential is greater on the surface of the Sun than on the surface of the Earth we observe a small but measurable gravitational redshift in the absorption lines of the Sun.

The second factor to shift the frequency is the Doppler effect. The Doppler effect arises when the observed star has

a radial velocity component in the direction of the Earth
based observer. If the Star is receding from the Earth then
the Doppler effect will create a redshift and conversely if
the star is approaching Earth the Doppler effect causes a
frequenct shift in the opposite direction - called a blueshift.
Redshift (Z) is given as the observed wavelength divided by
the emitted wave length less one, thus:-

$$Z = \gamma_o/\gamma_e - 1. \tag{12.1}$$

γ_o and γ_e are the wavelengths of the photon at observation
and at emission respectively.

The many absorption lines in the emitted spectrum of a
star act as frequency markers so that the degree of red or
blue shift can be very accurately determined. After allowing
for gravitational redshift the remaining Doppler shift enables
the radial velocity of the star to be accurately determined,
but on the assumption that there is no additional factor
causing a spectrum shift of the received light.

It was found by measures of the Doppler shift that nearby
stars in our galaxy, the Milky Way, showed various degrees
of red or blueshift, indicating radial velocities of up to 50
km/s for most stars. More powerful telescopes eventually
made it possible to measure the Doppler shift of galaxies
at far greater distances. In 1912 Vesto Slipher managed to
measure the radial velocity of the Andromeda galaxy which
he found to be 300 km/s towards Earth. Slipher also measure the radial velocity of the Sombrero galaxy which was
moving away from Earth at 1000 km/s. By 1917 Slipher
had measured the velocities of 25 galaxies, 21 of which were
moving away from Earth. The fact that the great majority
of galaxies were moving away from Earth was a considerable
mystery at the time.

In the 1920s Edwin Hubble, with the aid of the Mount

Wilson 100 inch telescope, determined to further investigate the radial velocity of galaxies. By 1929 he had measured the redshifts of 46 galaxies out to a distance of 7 million light years. In almost every case these galaxies were receding. Furthermore, when the recession velocity was plotted on a graph against galaxy distance their appeared to be a possible relationship between the two - generally the further the galaxy the faster the recession velocity. However this relationship was not at all certain at the distances measured as there was considerable variation in the recession velocities from the value predicted from the distance.

By 1931 Hubble had measured more galaxies at distances out to 100 million light years. The resultant graph now plainly showed a relationship between redshift (recession velocity) and distance. Unmistakably galaxies were receding from Earth at velocities which were proportional to their distance from us. The ratio of galactic velocity divided by galactic distance was later given the name of the Hubble constant, represented by H_0.

Over the years since 1931 the value of the Hubble constant has been considerably adjusted. The modification has particularly resulted from a more accurate measure of galactic distances which, in 1931, had been considerably underestimated. The current accepted value of the Hubble constant is approx. 70 km/s per mega parsec. Put in another way the distance to a galaxy increases by 1 part in 13.8 billion per year.

The Accepted Explanation of Galactic Redshift
It is not thought that the Hubble recession velocities (the Hubble flow) are actual velocities through Space. Establishment physicists and Aether theorists both agree that the Hubble recession velocities are a consequence of the expan-

sion of Space itself, albeit by radically different mechanisms.

As previously stated the frequency of the radiation from stars in distant galaxies occurs at exactly the same frequency as the emissions from local nearby stars. But over the time of the journey at the speed of light from the distant galaxy to Earth, Space expands at the Hubble rate of 1 part in 13.8 billion per year. The wavelength of starlight is therefore exposed to the expansion of Space throughout its journey. The resultant increase in wavelength causes the redshift observed by Hubble. For example light from a galaxy 100 million light years away has taken that time to reach Earth. Thus the wavelength will be increased by 1 part in 138; a not insignificant amount.

It might be thought from Hubble's observations that Space is expanding away from Earth rather as though Earth was at the centre of the Universe - but this is not so. On a large scale Space is expanding equally in all directions and thus is receding to an equal degree from every point of the Universe. An observer will see exactly the same recession effect no matter where he is within the Universe as on a large scale Space is considered to be isotropic and homogenous, which implies that there are no boundaries to the Universe.

The difference in the observed recession velocities from the Hubble velocity ($V = H_0 d$) as predicted from the measured distance, are considered to be velocities *through* the expanding Space. These through Space velocities - more obvious in nearby objects where the Hubble velocity is small - are called Peculiar Velocities. Although the Doppler-redshift method can only measure the radial component of the peculiar velocity it is reasonable to assume that the true direction of peculiar velocity is randomly orientated with respect to the direction of Earth. Thus the true peculiar velocity is generally higher than the measured radial peculiar velocity.

It is of consequence to realize that the establishment explanation of galactic redshift gives Space a form of physical entity, such as is possessed by the Aether, through which and relative to which the heavenly bodies are deemed to move with a through Space peculiar velocity.

The Big Bang Theory of Space Expansion

The expansion of the Universe moving forwards in time means that, looking backwards in time, the Universe contracts. The conclusion of modern physics is that this contraction continues back to an event prior to which neither Space nor Time existed. Their theory is that all the present day energy and mass of the Universe must have originated from that single point. The explosion of that point into our expanding and evolving Universe is called the Big Bang which is believed to have occurred around 14 billion years ago. The theory of that explosion, expansion and evolution of the Universe is called the Big Bang Theory.

The Big Bang theory of the expansion of the Universe relies upon the assumption that the total energy of the Universe remains a constant throughout the life of the Universe. The total energy is given by the sum of the total kinetic energy of all the matter in the universe and the total potential energy due to a universal gravitational attraction towards a centre of the Universe. The balance of the kinetic energy to the gravitational energy determines whether the Universe will continue to expand forever or will fall back upon itself as the gravitational pull eventually overcomes the expansion velocity and the Universe falls back into a scenario called the Big Crunch. But whatever the final outcome of the Universe the Big Bang theory requires that the expansion velocity of the Universe must be decelerating as every piece of matter in the universe continuously gravitationally attracts each other.

Thus the Big Bang theory cannot predict the acceleration of cosmic expansion - which is what is observed - without modification.

In order to fit the Big Bang theory to predict cosmic acceleration a new force was introduced ad-hoc by Einstein, called the Cosmological Constant. This invention is considered to be a form of Space energy which creates an expansion force upon Space. This new force overcomes the inward pull of gravity where gravity is weak at great distances from galaxies to create an accelerating expansion of Space. This postulated Space Energy is called Dark Energy.

It is accepted by Big Bang theorists that the peculiar velocity and the cosmic recession velocity of stars and galaxies are different forms of velocity. Yet the Big Bang theory assumes that cosmic recession velocity acts in a similar manner to peculiar velocity in causing the velocity effects associated with peculiar velocity. In particular cosmic recession velocity is falsely assumed to create kinetic energy in matter.

The modern Big Bang theory requires three separate ad-hoc mechanisms to generate Space expansion. Firstly there is the initial Big Bang which generates mass and energy from nothing and imparts an outward kinetic velocity to the Universe. This is quickly followed by a postulated period of hyper-inflation of the Universe created in order to overcome an otherwise fatal flaw in the Big Bang theory. The required rate of hyper-inflation is such that the outer edge of the Universe expanded at the incredible rate of 3×10^{41} times the speed of light. No explanation is given of the causes of hyper-inflation, why it started or why it later stopped.
Then there is the Cosmological Constant.

The obvious thought is that where a theory requires three separate ad-hoc mechanisms to explain the single phenomenon of cosmic expansion then it is highly probable that neither mechanism is correct.

The Aether Expansion Hypothesis

If it is accepted that Space is an Aether then the question of Cosmic Expansion must be addressed within that context. As the distance between any two points in Space is observed to increase over time then it follows that there must exist more Aether substance in a volume described by various markers in Space than at a previous time. Thus new Aether must be created by some mechanism.

It has previously been postulated as the basis for the Aether Theory of Gravity that mass generates new Aether. As the average density of mass in the Universe is very small it would appear that new Aether generated by mass alone might only account for a small fraction of Universal cosmic expansion. But it should be remembered that mass is not a special substance but merely Aether with complex high magnitude electric potentials superimposed. Hence mass generated Aether originates from the internal Aether of mass particles, and so the process of new Aether generation derives from the Aether itself. The assumption is therefore made that Aether everywhere generates new Aether. But the rate of new Aether generation within matter is a great magnification of the process occuring in empty Space.

The total effect of mass generated new Aether is likely to have significance within high mass density formations such as galaxies, etc. but has little significance in the outer regions of Space where the major contribution to Cosmic Expansion must come from Aether generated new Aether. The total rate of new Aether generation is one part in 4.6 billion per year, based upon the current accepted value of the Hubble constant.

It follows that new Aether must itself in turn generate

more new Aether. Thus the expansion of Space is a geometric one.

Geometric Space Expansion

The postulate of the geometric expansion of Space is mathematically described thus:-

$$V = V_0 e^{tH_V} \tag{12.2}$$

where H_V is the Hubble expansion constant for volume and $H_V = 3H_L = 3H_0$. Consequently:-

$$s = s_0 e^{tH_L} \tag{12.3}$$

where s is Space (proper) distance. Also:-

$$v = ds/dt = s_0 H_L e^{tH_L} \tag{12.4}$$

and:-

$$acc = d^2 s/dt^2 = s_0 H_L^2 e^{tH_L} \tag{12.5}$$

At the present time:-

$$s = s_0$$

$$v_0 = sH_L \tag{12.6}$$

and

$$acc_0 = sH_L^2 \tag{12.7}$$

Eqn.(12.4) describes the velocity of Cosmic expansion where $v_0 = cZ$ (it will be shown that this is true for all values of Z).

Eqn.(12.6) describes the current rate of the Hubble flow.

Cosmic Expansion v. Trans-Aether Velocity

Cosmic expansion velocity and trans-Aether or peculiar velocity are two entirely different forms of velocity. Trans-Aether velocity is velocity through the Aether and is identical to the velocities that we experience around us, notwithstanding that we must measure these velocities relative to Earth which has its own Aether velocity.

It is trans-Aether velocities which cause the velocity effects of length contraction, time dilation, mass increase and kinetic energy, etc. Trans-Aether velocity is limited to the local speed of light. On the other hand cosmic recession velocity is caused by the generation of new Aether. When the distance between two separated space points increases then it may not be unreasonable to interpret that as a velocity between those two points, but it is more direct to simply consider the effect as Space expansion. Cosmic expansion velocity does not cause the velocity effects associated with trans-Aether velocity.

There is therefore no theoretical limit to cosmic recession velocity as there is with trans-Aether velocity. Cosmic recession velocities may exceed the speed of light to any degree.

Redshift, Proper Distance and Apparent Magnitude

Consider the light received on Earth from a star situated in a distant galaxy. We are concerned here only with the effects of cosmic expansion so the trans-Aether or peculiar velocities are ignored. During the lifetime of the emitted photons their wavelengths will be subject to the effects of cosmic expansion such that the wavelength is affected directly by eqn.(12.3). Thus:-

$$\gamma_o = \gamma_e e^{TH_L}$$

where T is the lifetime of the photon. The duration T is

given by s_{LP}/c where s_{LP} is the length of the path of the photon.

The redshift is given by:-

$$Z = (\gamma_o - \gamma_e)/\gamma_e = (e^{TH_L} - 1) \qquad (12.8)$$

As each photon moves towards Earth through Space both the distance behind it to the source galaxy and the Space distance in front to Earth are expanding according to eqn. 12.3. Each segment of distance δs behind the photon has expanded as a function of the time since the photon passed that segment. The sum of all these segments, when the photon strikes Earth, is the proper distance, s_p, between Earth and the galaxy. Therefore:-

$$s_p = c \int_0^T e^{tH_L} dt = c/H_L \left[e^{tH_L} \right]_0^T$$

$$s_p = c/H_L \left[e^{TH_L} - 1 \right]$$

Substituting from eqn.(12.8):-

$$s_p = cZ/H_L \qquad (12.9)$$

And substituting from eqn.(12.6)

$$v = cZ \qquad (12.10)$$

Eqn.(12.9) is Hubble's equation.

Redshifts of $Z = 7$ have been observed which equates to a distance of 100 billion light years if these redshifts are entirely due to cosmic expansion. But it is always possible that very intense gravitational fields may play a part in observed redshifts.

Eqn.(12.10) is the derived form of Hubble's equation.

It can be seen that both eqns.(12.9) and (12.10) are linear with respect to Z irrespective of its value.

Because the relative magnitude of a star may be related to its proper distance s_p, eqn.(12.9) gives a relationship between relative magnitude and the log. of redshift.

Two effects separate to the cosmic expansion effect diminish light intensity at high redshifts. The first effect is due to the loss of energy from each photon directly due to the redshift ($E = h_e/\gamma$). The second effect arises from the cosmic expansion effect as less photons are received per second than are transmitted. The combined effect is to diminish the light flux received and hence to increase the observed magnitude by the factor $5log(1 + Z)$. This diminution must be allowed for in calculating distance from the apparent magnitude.

Cosmic Acceleration

As observation techniques improved it was possible to measure ever larger redshifts at even larger distances. As a consequence it became necessary to adapt, modify and invent entirely new techniques in order to measure the greater and greater distances involved. An obvious difficulty of measuring at great distance is the further away the star the fainter it appears when observed on Earth.

Among the brightest objects in the heavens are supernovae. Supernovae are the result of the explosion of a star, so they are only transient in their appearance, but as a result of their intense brightness they can be seen at very great distances. It was realized that one particular type of supernova, Type 1a, gave a peak brightness which was constant within 0.2 magnitudes and so could be employed as a 'standard candle' with a known absolute magnitude of -19.6. The distance to the supernova can then readily be calculated from the apparent magnitude as seen on Earth and the known absolute magnitude.

The Type 1a supernovae enabled redshifts to be measured greater than $Z = 1$ with galactic distances out to several billion light years. Of course the supernovae explosions that we see today actually occurred several billion years ago. Hence the Type 1a supernovae enable us to look back at the expansion history of the Universe over that period. It was fully expected that the expansion of the Universe would be observed to be decelerating due to the inward pull of gravity emanating from all the mass of the Universe.

Actual observations at high redshifts show that objects are dimmer than expected from the distance calculated from their redshifts. There may be several reasons for this dimming but the current establishment belief is that distance is not linear with redshift and the expansion of the Universe is accelerating. Of course, acceleration is intrinsic to the Aether theory of Space Expansion.

The Implications of Geometric Space Expansion

According to the postulate and the value of the Hubble constant the volume of the Universe doubles every 3.2 billion years $(ln2/H_V)$ going forward in time and halves over the same duration going backwards in time. But as we consider the Universe to be infinite in size then it will always be infinite in size at whatever point in time we take - in the past or the future. Consequently, under the geometric expansion hypothesis, the Universe does not have a specific start point and so does not have an age.

In an expanding Space the big question is that of matter density, for if the total amount of matter in the Universe remains constant while Space continuously expands then the average density of matter must continue to fall. Conversely, going back in time, the density of matter would increase for ever. However, if Universal matter density remained constant over time then new matter must be created at the same rate as the creation of new Aether. The problem with this concept lies in understanding a mechanism of continuous matter creation - a problem which has not yet been solved. But as matter does exist in the Universe, then at some time and by some mechanism or other that matter was created - whether in a single instant or continuously is just one part of HOW.

The Big Bang theory of matter creation suggests that matter was created instantly from nothing - which seems rather implausible. On the other-hand a theory of constant matter creation does have the whole Matter, Energy and Space of the Universe as a resource from which, by some means, new matter might be constructed.

Chapter 13

Appendices

Appendix 1
The Bending of a Light Ray Grazing the Sun

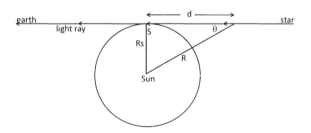

Figure 13.1: Light Ray grazing the Sun

Consider a light ray emitted by a distant star which passes the surface of the Sun at near zero height (the grazing point) and then passes on to infinity as depicted in Fig (13.1). The light ray passes through the speed of light gradient caused by the Sun which accelerates the light ray towards the Sun as a consequence of the equation:-

$$acc_r = sin\theta dc_L/dR \qquad (13.1)$$

The bending effect is at its greatest at the surface of the Sun where the speed of light gradient is a maximum. When the light pulse is some distance from the grazing point the speed of light gradient is diminished, and furthermore is at an increasing angle to the light ray path further diminishing its bending effect on the light ray.

The angle of bending of the ray from infinity in one direction to infinity in the other direction can be calculated from Fig (13.1).

Combining eqns. (13.1) with

$$c_L - c = -Amc/R \qquad (13.2)$$

gives:-

$$acc_x = Ac^2 m sin\theta/R^2$$

but $sin\theta = R_s/R$ thus:-

$$acc_x = Ac^2 m R_s/R^3$$

Now

$$R^2 = (R_s^2 + d^2)$$

so:-

$$acc_x = Amc^2 R_s/(R_s^2 + d^2)^{3/2} \qquad (13.3)$$

Fig.(13.2) describes the path of the light ray and the deflection caused by an acceleration in the x direction normal to the path of the ray.

Now if $\delta x \lll \delta d$ then $\delta\theta = \delta x/\delta d$.

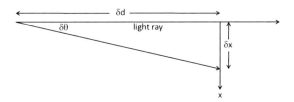

Figure 13.2: The Bending of a Light Ray

An acceleration in the x direction is given by $\delta x/\delta t^2$. Also $\delta t = \delta d/c$ therefore:-

$$\delta\theta = acc_x \delta d/c^2$$

substituting for acc_x from eqn. 13.3 gives:-

$$d\theta = \frac{Amr_s}{(R_s^2 + d^2)^{3/2}}\delta d$$

Integrating between d $= 0$ and infinity gives:-

$$\theta = Am/R_s$$

As $m = 1.47 kms$ and $R_s = 697 \times 10^3 kms$ then $\theta = 0.44A$ arc secs.

The total deflection of the light ray from infinity to infinity is twice the above figure at $0.88A$ arc secs. Eddington's observations, and subsequent more accurate ones, observed the actual deflection to be 1.75 arc secs.
Thus the value of A in eqn. (13.2) is determined to be 2 and so the equation may now be written as:-

$$c_L = c(1 - 2m/r)$$

Appendix 2
Increased Distance/Time through a Gravitational Field

As light slows in passing through a gravitational field it fol-

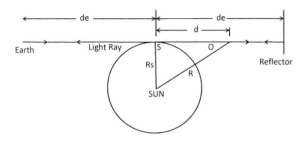

Figure 13.3: Radar Reflection grazing the Sun

lows that the time for a light ray or radio wave to reach a distant reflector and return to the sender will increase if a gravitational field is placed in the path of the ray where no field previously existed.

A radar signal is emitted from Earth such that it just grazes the Sun's surface on its way to a distant reflecting object (planet or space-craft) on the far side of the Sun. The distance from Earth to the reflecting object has been accurately predicted from previous knowledge of its orbit on the assumption that there is no intervening gravitational field. The distance to the reflector from Earth is next measured by timing the out and return journey of the radar pulse. The difference between the predicted and the measured distance is expected to be about $240\mu secs$ at the speed of light.

The calculation of the effect of the Sun's gravitational field is as follows:- The local speed of light $c_L = c\left(1 - \frac{2m}{R}\right)$ where $R = \sqrt{R_s^2 + d^2}$

$$\therefore \delta d/c = \left(1 - \frac{2m}{\sqrt{R_s^2 + d^2}}\right)\delta t. \qquad (13.1)$$

Hence

$$\delta d/c = \left[1 - \frac{2m}{c^2\left(\frac{R_s^2}{c^2} + t^2\right)^{1/2}}\right]\delta t$$

$$\therefore t_S = \left[t - \frac{2m}{c}\ln\frac{\left(t + \sqrt{t^2 + \frac{R_s^2}{c^2}}\right)}{R_s/c}\right]_T^0$$

t_S is the time for a light ray to cover the distance d_E or d_B with the sun present while t is the time when the sun is absent. T is the time for a light ray to pass from S to either point B or E. In the case of the Earth (E), $T = 500\mu S$.

The factor $R_S/c = 2.32$s.

$$\therefore t_S = t - \frac{2m}{c}\left(\ln 1 - \ln\frac{1000}{2.32}\right). \qquad (13.2)$$

The factor $2m/c = 9.8\mu s$ \qquad\qquad Hence $t_S - t = 59.5\mu S$

If the reflector B is the same distance from the Sun as the Earth then the round trip difference is $4(t_s - t) = 238\mu S$. Therefore the distance measurement through Space from Earth to the reflector has increased by 35.7 km, caused by the gravitational field of the Sun.

The predicted extra distance agrees accurately with several observations made from Earth to Venus by Shapiro et al, 1972. Once more confirming that the gravitational constant for light, G_L is twice that of matter, $(A = 2)$.

Appendix 3
The Precession of Mercury

It had been observed that the orbit of Mercury precesses 42 arc secs per century more than Newtonian gravity could explain.

The precession effect occurs in all orbits whether circular or elliptical but naturally the precession cannot be detected in a circular or near-circular orbit.

The precession effect is most simply explained with a circular orbit where:-

$$V^2/R = mc^2/r^2 \qquad (13.3)$$

The RH side of eqn. 13.3 describes the gravitational acceleration and the LH side describes the rate of bending of the path of the planet towards the Sun in order to describe an orbit of effective radius R. r is the separation of the planet and the Sun in sub-Aether units.

V may be given in terms of a fraction of the local speed of light, thus:-

$$V^2(1 - 2m/r)^2/R = m/r^2$$

The radii r and R are slightly different from each other. Where R is large relative to m which is the case in the solar system ($m = 1.47 kms$ for the Sun) we have:-

$$V^2(1 - 4m/r) = mR/r^2$$

Therefore the effective radius of the orbit $R = r(1 - 4m/r)$ when m/r is very small. The orbit length therefore exceeds the nominal orbit length of $2\pi r$ by $8\pi m$ or $36.9 km$ in the solar system irrespective of orbit radius. For the planet Mercury, $r = 58 \times 10^6 kms$, hence the precession is 0.1313 arc secs/orbit or 55 arc secs per century, which roughly accounts for the observed anomaly.

It can be seen from the factor $(1 - 4m/r)$ that strange effects would occur if the orbit radius r could approach the gravitational radius of the mass. This does not occur with the Sun as the radius of the Sun is $700,000kms$ and its gravitational radius is only $1.47kms$. However neutron stars are very much denser than the Sun so it may be possible that orbits could approach the gravitational radius.

Appendix 4
The Lorenz Velocity Transform
A body moving through the Aether at instantaneous velocity u may also be observed from an IRF which itself is moving through the Aether at the constant velocity v. The velocity of the body when observed from the IRF is found to be u'. Both u and u' are instantaneous velocities so the body may be accelerating. This particular point is very important as all real bodies are accelerating to one degree or another. The Lorenz Velocity Transforms relate the three velocities; u, u' and v.

The Lorenz Transforms may be written in the form:-

$$x = \gamma(x' + vt') \qquad (13.4)$$
$$t = \gamma(t' + vx'/c^2) \qquad (13.5)$$
$$y = y' \qquad (13.6)$$
$$z = z' \qquad (13.7)$$

Taking differentials and dividing eqn. (13.4) by eqn. (13.5) gives:-

$$dx/dt = (dx' + vdt')/(dt' + dx'v/c^2)$$

Dividing throughout by dt' gives:-

$$u_x = (u'_x + v)/(1 + u'_x v/c^2) \qquad (13.8)$$

which is the velocity transform equation for the x axis.

The velocity equations for the y and z axes are similarly found to be:-

$$u_y = u'_y/\gamma(1 + u'_x v/c^2) \qquad (13.9)$$

$$u_z = u'_z/\gamma(1 + u'_x v/c^2) \qquad (13.10)$$

The final Lorenz Transform equation can be obtained by combining the Transform equations for the x, y and z velocity components using the following two equations;

$$u = (u_x^2 + u_y^2 + u_z^2) \qquad \text{and} \qquad u' = (u_x'^2 + u_y'^2 + u_z'^2)$$

From eqns. (13.8), (13.9) and (13.10) we now obtain by mathematical manipulation (once again standard text book stuff)

$$\gamma_u/\gamma_u' = \gamma_v(1 + u_x'v/c^2) \tag{13.11}$$

Equation (13.11) relates the three Lorenz functions for velocities u, u' and v and is called the Lorenz Velocity Transform equation.